POWER SUPPLY DEVICES

and Systems of Relay Protection

POWER SUPPLY DEVICES

and Systems of Relay Protection

VLADIMIR GUREVICH

CRC Press
Taylor & Francis Group
Boca Raton London New York

CRC Press is an imprint of the
Taylor & Francis Group, an **informa** business

CRC Press
Taylor & Francis Group
6000 Broken Sound Parkway NW, Suite 300
Boca Raton, FL 33487-2742

First issued in paperback 2017

© 2014 by Taylor & Francis Group, LLC
CRC Press is an imprint of Taylor & Francis Group, an Informa business

No claim to original U.S. Government works

Version Date: 20130412

ISBN 13: 978-1-4665-8379-5 (hbk)
ISBN 13: 978-1-138-07503-0 (pbk)

Library of Congress Cataloging-in-Publication Data

Gurevich, Vladimir, 1956-
  Power supply devices and systems of relay protection / Vladimir Gurevich.
    pages cm
  Includes bibliographical references and index.
  ISBN 978-1-4665-8379-5 (hardback)
  1. Power electronics. 2. Electric power systems--Equipment and supplies. 3. Electric power supplies to apparatus. 4. Protective relays. I. Title.

TK7881.15.G844 2013
621.381'044--dc23   2013001997

Visit the Taylor & Francis Web site at
http://www.taylorandfrancis.com

and the CRC Press Web site at
http://www.crcpress.com

Contents

Preface

Development of modern machinery and transition to microprocessor-based elements is accompanied by constant sophistication. Modern digital protection relay (DPR) devices, unlike old electromechanical relays, require power supplies. Quality and reliability of relay protection (RP) power supply systems and devices affect RP's ability to perform its functions. The power supply of RP starts at a transformer for a substation's own needs and ends at a built-in power supply of the DPR, including a system of auxiliary current, battery chargers, accumulator batteries, uninterruptable power supply (UPS), auxiliary systems of monitoring of insulation, and integrity of auxiliary current circuits.

All these systems and devices are connected by many links and represent an integral organism, where a failure of one organ can result in a serious "illness" of the whole organism. For example, routine work to find a location of insulation breakage in a 220 V DC system by means of a standard device, which has been done multiple times and with which all electricians are so familiar, can suddenly result in the switching off of a power transformer and a full range of overhead transmission lines at 160 kV, redistribution of the load to other lines, their overload, and, finally, the breakdown of the whole energy system. Why?

There is another problem: When working at a substation, which should be completely insulated from the ground, an electrician accidentally grounds one of the DC pole terminals. As a result, internal power supplies of dozens of DPRs are broken. Again, we need to ask: why?

Consider a simpler situation: We need to select an accumulator battery for a substation. One supplier offers GroE batteries and another offers OGi batteries; both of them praise their products. Moreover, based on advertising materials, both types are equal. How should we act in this situation? How should we select a battery charger, if we do not know which types are available and what the differences between them are? Do we need an active harmonics filter, which is strictly recommended by the seller of equipment as a solution to all problems? Is a UPS that distorts current consumed from a general supply line that has a level of coefficient of nonlinear distortions as high as 40%?

The answers to these questions are rather difficult and require that the staff maintaining and using AC and DC auxiliary power supply systems possess a certain level of knowledge. The lack or shortage of this knowledge not only interferes with maintaining RP power supply systems at an adequate level, but also can sometimes be a source of serious breakdowns and accidents.

The author of this book describes auxiliary power supply systems and devices in detail: built-in DPR power supplies, battery chargers, accumulator batteries, UPS, and characteristic features of auxiliary DC systems at substations and power plants. The author also discusses specific problems of RP power supply systems and devices, which are almost unknown and are not described in the technical literature because they are unclear.

It is also very important to know how to solve problems. That is why the description of technical problems is accompanied by corresponding solutions. The author also has tried to solve the problem of staff who service RP power supply systems not having enough knowledge in the area of electronics, which makes it difficult for them to work with this equipment on a daily basis. This task is addressed by describing the backgrounds of electronics and primary elements of the system in the first chapter of the book: transistors, thyristors, optocouplers, logic elements, and relays.

The book is meant for engineers and technicians who use AC and DC auxiliary power systems of power plants and substations, as well as relay protection systems. The book may be useful for teachers and students of corresponding disciplines at vocational schools and higher education institutions.

Please send your remarks about the book to the author: vladimir.gurevich@gmail.com

The Author

Vladimir I. Gurevich was born in Kharkov, Ukraine, in 1956. He received an MSEE degree (1978) from Kharkov Technical University and a PhD degree (1986) from Kharkov National Polytechnic University. His employment experience includes teacher, assistant professor, and associate professor at Kharkov Technical University; and chief engineer and director of Inventor, Ltd. In 1994, he arrived in Israel and works today at Israel Electric Corporation as an engineer specialist and head of a section of the Central Electric Laboratory.

He is the author of more than 150 professional papers and 9 books and holder of nearly 120 patents in the field of electrical engineering and power electronics. In 2006, he became an honorable professor with Kharkov Technical University. Since 2007, he has served as an expert with the TC-94 Committee of the International Electrotechnical Commission (IEC).

The author has published the following books with Taylor & Francis:

- *Protection Devices & Systems for High Voltage Applications*
- *Electrical Relays: Principles and Applications*
- *Electronic Devices on Discrete Components for Industrial and Power Engineering*
- *Digital Protective Relays: Problems and Solutions*

1

Basic Components

1.1 Semiconducting Materials and p–n Junctions

As is known, all substances, depending on their electroconductivity, are divided into three groups: conductors (usually metals) with a resistance of 10^{-6}–10^{-3} Ωcm, dielectrics with a resistance of 10^9–10^{20} Ωcm, and semiconductors (many native-grown and artificial crystals) covering an enormous intermediate range of values of specific electrical resistance.

The main peculiarity of crystal substances is typical, well-ordered atomic packing into peculiar blocks—crystals. Each crystal has several flat symmetric surfaces and its internal structure is determined by the regular positional relationship of its atoms, which is called the lattice. Both in appearance and in structure, any crystal is like any other crystal of the same given substance. Crystals of various substances are different. For example, a crystal of table salt has the form of a cube. A single crystal may be quite large in size or so small that it can only be seen with the help of a microscope. Substances having no crystal structure are called amorphous. For example, glass is amorphous in contrast to quartz, which has a crystal structure.

Among the semiconductors that are now used in electronics, one should point out germanium, silicon, selenium, copper-oxide, copper sulfide, cadmium sulfide, gallium arsenide, and carborundum. To produce semiconductors, two elements are mostly used: germanium and silicon.

In order to understand the processes taking place in semiconductors, it is necessary to consider phenomena in the crystal structure of semiconductor materials, which occur when their atoms are held in a strictly determined relative position to each other due to weakly bound electrons on their external shells. Such electrons, together with electrons of neighboring atoms, form *valence bonds* between the atoms. Electrons taking part in such bonds are called *valence electrons*. In absolutely pure germanium or silicon at very low temperatures, there are no free electrons capable of creating electric current, because under such circumstances all four valence electrons of the external shells of each atom that can take part in the process of charge transfer are too strongly held by the valence bounds. That is why this substance is an insulator (dielectric) in the full sense of the word: It does not let electric current pass at all.

When the temperature is increased, due to the thermal motion some valence electrons detach from their bonds and can move along the crystal lattice. Such electrons are called *free electrons*. The valence bond from which the electron is detached is called a *hole*. It possesses properties of a positive electric charge, in contrast to the electron, which has a negative electric charge. The higher the temperature is, the more free electrons that are capable of moving along the lattice and the higher the conductivity of the substance.

Moving along the crystal lattice, free electrons may run across holes—valence bonds missing some electrons—and fill up these bonds. Such a phenomenon is called *recombination*. At normal temperatures in the semiconductor material, free electrons occur constantly, and recombination of electrons and holes takes place.

If a piece of semiconductor material is put into an electric field by applying a positive or negative terminal to its ends, for instance, electrons will move through the lattice toward the positive electrode and holes to the negative one. The conductivity of a semiconductor can be enhanced considerably by applying specially selected admixtures (metal or nonmetal) to it. In the lattice, the atoms of these admixtures will replace some of the atoms of the semiconductors. Let us remind ourselves that external shells of atoms of germanium and silicon contain four valence electrons and that electrons can only be taken from the external shell of the atom. In their turn, the electrons can be added only to the external shell, and the maximum number of electrons on the external shell is eight.

When an atom of the admixture has more valence electrons than required for valence bonds with neighboring atoms of the semiconductor, additional free electrons capable of moving along the lattice occur on it. As a result the electroconductivity of the semiconductor increases. As germanium and silicon belong to the fourth group of the periodic table of chemical elements, donors for them may be elements of the fifth group, which have five electrons on the external shell of atoms. Phosphorus, arsenic, and stibium belong to such donors (*donor admixture*).

If admixture atoms have fewer electrons than needed for valence bonds with surrounding semiconductor atoms, some of these bonds turn out to be vacant and holes will occur in them. Admixtures of this kind are called *p*-type ones because they absorb (accept) free electrons. For germanium and silicon, *p*-type admixtures are elements from the third group of the periodic table of chemical elements, the external shells of atoms that contain three valence electrons. Boron, aluminum, gallium, and indium can be considered *p*-type admixtures (*accepter admixture*).

In the crystal structure of a pure semiconductor, all valence bonds of neighboring atoms turn out to be fully filled, and occurrence of free electrons and holes can be caused only by deformation of the lattice arising from thermal or other radiation. Because of this, conductivity of a pure semiconductor is quite low under normal conditions.

If some donor admixture is injected, the four electrons of the admixture, together with the same number in the filled valence, bond with the latter. The fifth electron of each admixture atom appears to be "excessive" or "redundant" and therefore can freely move along the lattice.

When an accepter admixture is injected, only three filled valence bonds are formed between each atom of the admixture and neighboring atoms of the semiconductor. One electron is lacking to fill up the fourth. This valence bond appears to be vacant. As a result, a hole occurs. Holes can move along the lattice like positive charges, but instead of an admixture atom, which has a fixed and permanent position in the crystal structure, the vacant valence bond moves.

It goes like this. An electron is known to be an elementary carrier of an electric charge. Affected by different causes, the electron can escape from the filled valence bond, leaving a hole that is a vacant valence bond and that *behaves like a positive charge equaling numerically the negative charge of the electron.* Affected by the attracting force of its positive charge, the electron of another atom near the hole may "jump" to the hole. At that point, recombination of the hole and the electron occurs, their charges are mutually neutralized, and the valence bond is filled. The hole in this place of the lattice of the semiconductor disappears.

In its turn a new hole, which has arisen in the valence bond from which the electron has escaped, may be filled with some other electron that has left a hole. Thus, moving of electrons in the lattice of the semiconductor with a *p*-type admixture and recombination of them with holes can be regarded as moving of holes. For better understanding, one may imagine a concert hall in which for some reason some seats in the first row turn out to be vacant. As spectators from the second row move to the vacant seats in the first row, their seats are taken by spectators of the third row, etc. One can say that, in some sense, vacant seats "move" to the last rows of the concert halls, although in fact all the stalls remain screwed to the floor. "Moving" of holes in the crystal is very much like "moving" of such vacant seats.

Semiconductors with electroconductivity enhanced, due to an excess of free electrons caused by admixture injection, are called semiconductors with *electron-conductivity* or, in short, *n-type semiconductors.* Semiconductors with electroconductivity influenced mostly by moving of holes are called *semiconductors with p-type conductivity* or just *p-type semiconductors.*

There are practically no semiconductors with only electronic or only *p*-type conductivity. In a semiconductor of *n*-type, electric current is partially caused by the moving of holes arising in its lattice because of electrons escaping from some valence bonds; in semiconductors of *p*-type, current is partially created by the moving of electrons. Because of this, it is better to define semiconductors of the *n*-type as semiconductors in which *the main current carriers are electrons* and semiconductors of the *p*-type as semiconductors in which *holes are the main current carriers.* Thus, a semiconductor belongs to this or that type, depending on what type of current carrier predominates in it. According to this, the other opposite charge carrier for any semiconductor of a given type is a *minor carrier.*

One should take into account that any semiconductor can be made a semiconductor of *n*- or *p*-type by putting certain admixtures into it. In order to obtain the required conductivity, it is enough to put in a very small amount of the admixture—about one atom of the admixture for 10 millions of atoms of the semiconductor. All of this imposes special requirements for the purification of the original semiconductor material and accuracy in dosage of admixture injection. One should also take into consideration that the speed of current carriers in a semiconductor is lower than in a metal conductor or in a vacuum. Moving of electrons is slowed down by obstacles on their way in the form of inhomogeneities in the crystal. Moving of holes is half as slow because they move due to jumping of electrons to vacant valence bounds. Mobility of electrons and holes in a semiconductor is increased when the temperature goes up. This leads to an increase of conductivity of the semiconductor.

The functioning of most semiconductors is based on the processes taking place in an intermediate layer formed in the semiconductor, at the boundary of the two zones with the conductivities of the two different types: *p* and *n*. The boundary is usually called the *p–n junction* or the *electron–hole junction,* in accordance with the main characteristics of the types of main charge carriers in the two adjoining zones of the semiconductor.

There are two types of *p–n* junctions: *planar* and *point junctions,* which are illustrated schematically in Figure 1.1. A planar junction is formed by moving a piece of the admixture—for instance. indium to the surface of the germanium—of *n*-type and further heating until the admixture is melted.

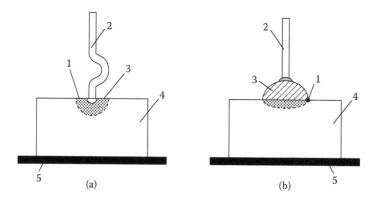

FIGURE 1.1
Construction of point (a) and planar (b) *p–n* junctions of the diode. 1—*p–n* junction; 2—wire terminal; 3—*p*-area; 4—crystal of *n*-type; 5—metal heel piece.

When a certain temperature is maintained for a certain period of time, there is diffusion of some admixture atoms to the plate of the semiconductor to a small depth, and a zone with conductivity opposite to that of the original semiconductor is formed. In the preceding case, it is *p*-type, for *n*-germanium.

Point junction results from tight electric contact of the thin metal conductor (wire), which is known to have electric conductivity, with the surface of the *p*-type semiconductor. This was the basic principle on which the first crystal detectors operated. To decrease dependence of diode properties on the position of the pointed end of the wire on the surface of the semiconductor and the clearance of its momentary surface point, junctions are formed by fusing the end of the thin metal wire to the surface of a semiconductor of the *n*-type. Fusion is carried the moment a short-term powerful pulse of electric current is applied. Affected by the heat formed for this short period of time, some electrons escape from atoms of the semiconductor near the contact point and leave holes. As a result of this, some small part of the *n*-type semiconductor in the immediate vicinity of the contact turns into a semiconductor of the *p*-type (area 3 on Figure 1.1a).

Each part of semiconductor material, taken separately (that is, before contacting), is neutral, since there is a balance of free and bound charges (Figure 1.2a). In the *n*-type area, concentration of free electrons is quite high and that of holes is quite low. In the *p*-type area, on the contrary, concentration of holes is high and that of electrons is low. Joining of semiconductors with different concentrations of main current carriers causes diffusion of these carriers through the junction layer of these materials: The main carriers of the *p*-type semiconductor—holes—diffuse to the *n*-type area because the concentration of holes in it is very low. And, vice versa, electrons from the *n*-type semiconductor, with a high concentration of them, diffuse to the *n*-type area, where there are few of them (Figure 1.2b).

On the boundary of the division of the two semiconductors, from each side a thin zone with conductivity opposite to that of the original semiconductor is formed. As a result, on the boundary (which is called a *p–n* junction), a space charge arises (the so-called potential barrier) that creates a diffusive electric field and prevents the main current carriers from flowing after balance has been achieved (Figure 1.2c).

Strongly pronounced dependence of electric conductivity of a *p–n* junction, from polarity of external voltage applied to it, is typical of the *p–n* junction. This can never be noticed in a semiconductor with the same conductivity. If voltage applied from the outside creates an electric field coinciding with a diffusive electric field, the junction will be blocked and current will not pass through it (Figure 1.3).

Moreover, moving of minor carriers becomes more intense, which causes enlargement of the blocking layer and lifting of the barrier for main carriers. In this case it is usually said that the junction *is reversely biased*. Moving of minor carriers causes a small current to pass through the blocked junction.

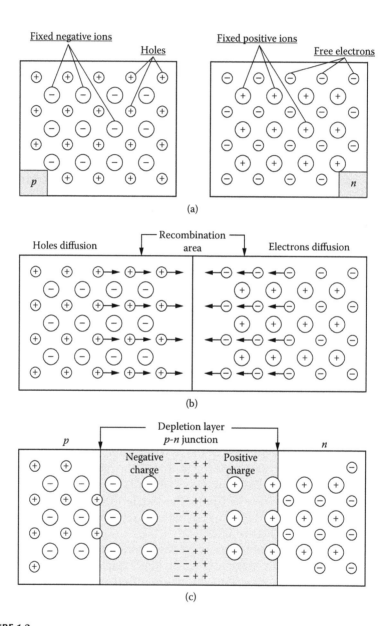

FIGURE 1.2
Formation of a blocking layer when semiconductors of different conductivity are connected.

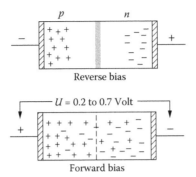

FIGURE 1.3
p–n Junction with reverse and forward bias.

This is the so-called *reverse current* of the diode, or *leakage current*. The smaller it is, the better the diode is.

When the polarity of the voltage applied to the junction is changed, the number of main charge carriers in the junction zone increases. They neutralize the space charge of the blocking layer by reducing its width and lowering the potential barrier that prevented the main carriers from mobbing through the junction. It is usually said that the junction is *forward biased*. The voltage required for overcoming the potential barrier in the forward direction is about 0.2 V for germanium diodes and 0.6–0.7 V for silicon ones.

To overcome the potential barrier in the reverse direction, tens and sometimes even thousands of volts are required.

If the barrier is passed over, irreversible destruction of the junction and its breakdown takes place, which is why threshold values of reverse voltage and forward current are indicated for junctions of different appliances.

Figure 1.4 illustrates an approximate volt-ampere characteristic of a single junction, which is dependent on current passing through it on the polarity

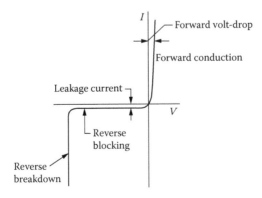

FIGURE 1.4
Volt-ampere characteristic of a single *p–n* junction (diode).

and external voltage applied to the junction. Currents of forward and reverse direction (up to the breakdown area) may differ by tens and hundreds of times. As a rule, planar junctions withstand higher voltages and currents than point ones, but do not work properly with high-frequency currents.

1.2 The Transistor's Principle

The idea of somehow using semiconductors had been tossed about before World War II, but knowledge about how they worked was scant, and manufacturing semiconductors was difficult. In 1945, however, the vice president for research at Bell Laboratories established a research group to look into the problem. The group was led by William Shockley and included Walter Brattain, John Bardeen, and others—physicists who had worked with quantum theory, especially in solids. The team was talented and worked well together.

In 1947 John Bardeen and Walter Brattain, with colleagues, created the first successful amplifying semiconductor device. They called it a transistor (from "transfer" and "resistor"). In 1950 Shockley made improvements to it that made it easier to manufacture. His original idea eventually led to the development of the silicon chip. Shockley, Bardeen, and Brattain won the 1956 Nobel Prize for the development of the transistor. It allowed electronic devices to be built smaller, lighter, and even more cheaply.

It can be seen in Figure 1.5 that a transistor contains two semiconductor diodes that are connected together and have a common area. Two utmost layers of the semiconductor (one of them is called an "emitter" and the other a "collector") have p-type conductivity with a high concentration of holes, and the intermediate layer (called a "base") has n-type conductivity with a low concentration of electrons. In electric circuits, low voltage is applied to the first (the emitter) p–n junction because the junction is connected in the forward (carrying) direction, and much higher voltage is applied to the second (the collector) junction, in the reverse (cutoff) direction. In other words, an emitter junction is forward biased and a collector junction is reverse biased.

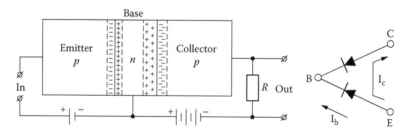

FIGURE 1.5
Circuit and the principle of operation of a transistor.

The collector junction remains blocked until there is no current in the emitter-base circuit. The resistance of the whole crystal (from the emitter to the collector) is very high. As soon as the input circuit (Figure 1.5) is closed, holes from the emitter seem to be injected (emitted) to the base and quickly saturate it (including the area adjacent to the collector). As the concentration of holes in the emitter is much higher than the concentration of electrons in the base, after recombination there are still many vacant holes in the base area, which is affected by the high voltage (a few or tens of volts) applied between the base and the collector, easily passing over the barrier layer between the base and the collector.

Increased concentration of holes in the cutoff collector junction causes the resistance of this junction to fall rapidly, and it begins to *conduct current in the reverse direction*. The high strength of the electric field in the "base–collector" junction results in a very high sensitivity of the resistance of this junction in the reverse (cutoff) state to a concentration of the holes in it.

That is why even a small number of holes injected from the emitter under the effect of weak input current can lead to sharp changes of conductivity of the whole structure and considerable current in the collector circuit.

The ratio of collector current to base current is called the *current amplification factor*. In low-power transistors, this amplification factor has values of tens and hundreds and, in power transistors, tens.

1.3 Some Types of Transistors

In the 1970s, transistor engineering developed very rapidly. Hundreds of types of transistors and new variants of them appeared (Figure 1.6). Among them appeared transistors with reverse conductivity or *n–p–n* transistors, as well as unijunction transistors. (Because it contains only one junction, such a transistor is sometimes called a two-base diode; see Figure 1.7.) This transistor contains one junction formed by welding a core made from *p*-material to a single-crystal wafer made from *n*-type material (silicon). The two outlets, serving as bases, are attached to the wafer. The core, placed asymmetrically with regard to the base, is called an emitter. Resistance between the bases is about a few thousand ohms. Usually, the base B_2 is biased in a positive direction from the base B_1. Application of positive voltage to the emitter causes strong current of the emitter (with insignificant voltage drop between the emitter E and the base B_1). One can observe the area of negative resistance (NR; see Figure 1.7) on the emitter characteristic of the transistor where the transistor is very rapidly enabled, operating like a switch.

In fact, modern transistors (Figure 1.8) are characterized by such a diversity of types that it is simply impossible to describe all of them in this book. Therefore, only a brief description of the most popular types of modern semiconductor devices and the relays based on them is presented here.

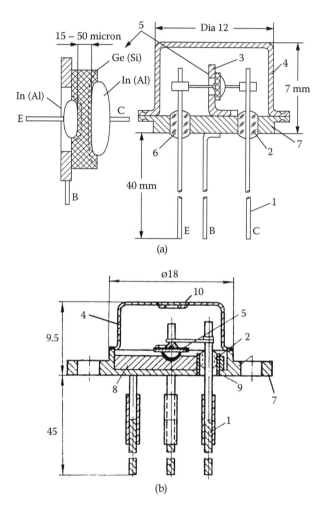

FIGURE 1.6
Transistors produced in the 1970s: (a) low power transistor; (b) power transistor. 1—outlets; 2 and 6—glass insulators; 3—crystal holder; 4—protection cover; 5—silicon (germanium) crystal; 7—flange; 8—copper heat sink; 9—Kovar bushing; 10—hole for gas removal after case welding and disk for sealing in. All measurements are in millimeters; ø = diameter.

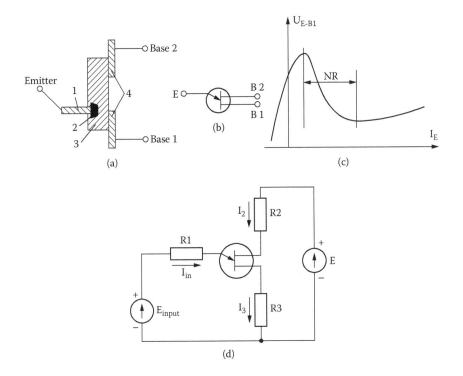

FIGURE 1.7
A unijunction transistor (or two-base diode) (a) construction: 1—*p*-type core; 2—*p*–*n* junction; 3—*n*-type plate; 4—ohmic contacts; (b) schematic diagram; (c) characteristic: NR—negative resistance area; (d) connection diagram.

In addition to the transistors described previously, which are called *bipolar junction transistors* or just "bipolar transistors" (Figure 1.9), so-called field effect transistors (FET; Figure 1.10) have become very popular recently. The first person to attempt to construct a field effect transistor in 1948 was, again, William Shockley. But it took many years of additional experiments to create a working FET with a control *p*–*n* junction called a "unitron" (Unipolar Transistor), in 1952.

Such a transistor was a semiconductor three-electrode device in which control of the current caused by the ordered motion of charge carriers of the same sign between two electrodes was carried out with the help of an electric field (that is why it is called "field") applied to the third electrode.

Electrodes between which working currents pass are called *source* and *drain* electrodes. The source electrode is the one through which carriers flow into the device. The third electrode is called a *gate*. Change of value of the working current in a unipolar transistor is carried out by changing the effective resistance of the current conducting area, the semiconductor material between the source and the drain called the *channel*. That change is made by increasing or decreasing area 5 (Figure 1.10). Increase of voltage of the initial

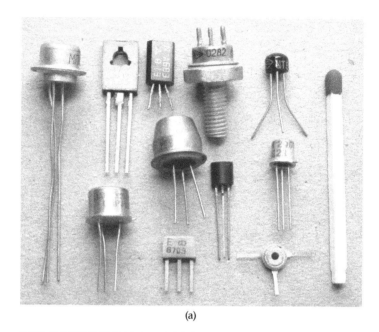

(a)

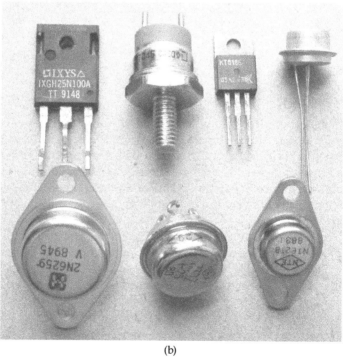

(b)

FIGURE 1.8
This is how modern low signal (a) and power (b).   *Continued*

(c)

FIGURE 1.8 (*Continued*)
(c) High power transistors.

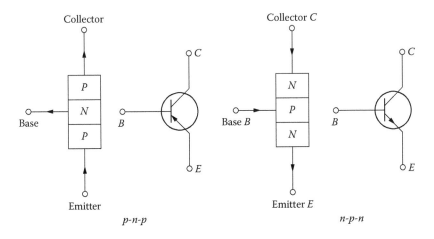

FIGURE 1.9
Structure and symbolic notation on the schemes of bipolar transistors of *p–n–p* and *n–p–n* types.

junction bias leads to expansion of the depletion layer. As a result, the rest area of the section of the conductive channel in the silicon decreases and the transistor is blocked, and vice versa. When the value of the blocking voltage on the gate decreases, the area (5) depleted by current carriers contracts and turns into a pointed wedge. At the same time, the section of the conductive channel increases and the transistor is enabled.

Depending on the type of the conductivity of semiconductor material of the channels, there are unipolar transistors with *p* and *n* channels. Because control of the working current of unipolar transistors is carried out with the help of a channel, they are also called *channel transistors.* The third name

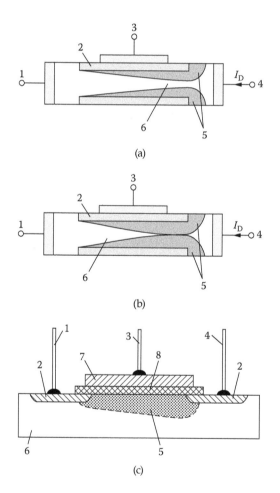

FIGURE 1.10
Simplified structure of FET and MOSFET transistors 1—source; 2—*n*-type admixture; 3—gate; 4—drain; 5—area consolidated by current carriers (depletion layer); 6—conductive channel in silicon of *p*-type; 7—metal; 8—silicon dioxide.

of the same semiconductor device—a *field transistor or field effect transistor*—points out that working current control is carried out by an electric field (voltage) instead of electric current as in a bipolar transistor. The latter peculiarity of unipolar transistors, which allows them to obtain very high input resistances estimated in tens and hundreds of megaohms, determined their most popular name: field transistors.

It should be noted that, apart from field transistors with $p–n$ junctions between the gate and the channel (FET), there are also field transistors with an insulated gate: metal oxide semiconductor FET transistors (MOSFETs). The latter were suggested by S. Hofstein and F. Heiman in 1963.

Field transistors with an insulated gate appeared as a result of searching for methods to further increase input resistance and frequency range extensions of field transistors with $p–n$ junctions. The distinguishing feature of such field transistors is that the junction biased in a reverse direction is replaced with a control structure MOSFET form. As shown in Figure 1.10, this device is based on a silicon monocrystal, in this case of p-type. The source and drain areas have conductivity opposite to the rest of the crystal that is of the n-type. The distance between the source and the drain is very small, usually about 1 μm. The semiconductor area between the source and the drain, which is capable of conducting current under certain conditions, is called a channel, as in the previous case (Figure 1.11).

In fact, the channel is an n-type area formed by diffusion of a small amount of the donor admixture to the crystal with p-type conductivity. The gate is a metal plate covering source and drain zones. It is isolated from the monocrystal by a dielectric layer only 0.1 μm thick. The film of silicon dioxide formed at this high temperature is used as a dielectric. Such film allows us to adjust the concentration of the main carriers in the channel area by changing both value and polarity of the gate voltage (Figure 1.12).

This is the major difference of MOSFET transistors, as opposed to field ones with $p–n$ junctions, which can only operate well *with blocking voltage of the gate*. The change of polarity of the bias voltage leads to junction unblocking and to a sharp reduction of the input resistance of the transistor.

The basic advantages of MOSFET transistors are as follows: First, there is an insulated gate allowing an increase in input resistance by at least 1,000 times in comparison with the input resistance of a field transistor with a $p–n$ junction. In fact, it can reach a billion MΩ. Second, gate and drain capacities become considerably lower and usually do not exceed 1–2 pF. Third is the

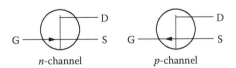

FIGURE 1.11
Symbolic notation of FET transistors with n- and p-channels: G—gate; S—source; D—drain.

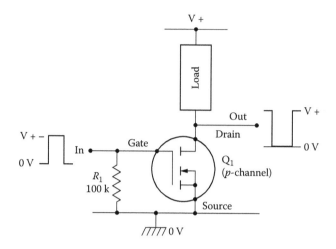

FIGURE 1.12
Symbolic notation and the circuit of a MOSFET transistor.

limiting frequency of MOSFET—transistors can reach 700–1000 MHz—at least 10 times higher than that of standard field transistors.

Attempts to combine in one switching device the advantages of bipolar and field transistors led to the invention of a compound structure in 1978, which was called a *pobistor* (Figure 1.13). The idea of a modular junction of crystals of bipolar and field transistors in the same case was employed by Mitsubishi Electric to create a powerful switching semiconductor module (Figure 1.14).

Further development of production technology of semiconductor devices allowed development of a single-crystal device with a complex structure with properties of a pobistor: an *insulated gate bipolar transistor* (IGBT). The IGBT is a device that combines the fast-acting features and high-power capabilities of the bipolar transistor with the voltage control features of the MOSFET gate. In simple terms, the collector–emitter characteristics are similar to those of

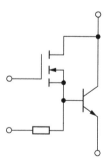

FIGURE 1.13
Compound structure: *pobistor.*

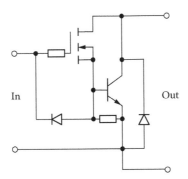

FIGURE 1.14
Scheme of a power switching module CASCADE-CD, with a working voltage of 1000 V and currents more than 100 A (Mitsubishi Electric).

the bipolar transistor, but the control features are those of the MOSFET. The equivalent circuit and the circuit symbol are illustrated in Figure 1.15. Such a transistor (Figure 1.16) has a higher switching power than FET and bipolar transistors and its operation speed is between that of FET and bipolar transistors. Unlike bipolar transistors, the IGBT does not operate well in the amplification mode and is designed for use in the switching (relay) mode as a powerful high speed switch.

The IGBT is enabled by a signal of positive (with regard to the emitter) polarity, with voltage not more than 20 V. It can be blocked with zero potential on the gate; however, with some types of loads, a signal of negative polarity on the gate may be required for reliable blocking (Figure 1.17).

Many companies produce special devices for IGBT control. They are made as separate integrated circuits or ready-to-use printed circuit cards, so-called *drivers* (Figure 1.18). Such drivers are universal as a rule and can be applied to

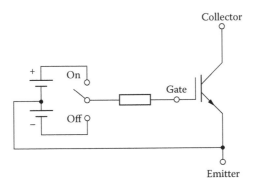

FIGURE 1.15
Insulated gate bipolar transistor (IGBT).

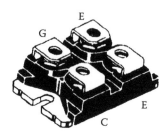

FIGURE 1.16

An IGBT IXDN75N120A produced by IXYS with a switched current up to 120 A and maximum voltage of up to 1200 V (dissipated power is 630 W). With such high parameters, the device is quite small in size: 38 × 25 × 12 mm.

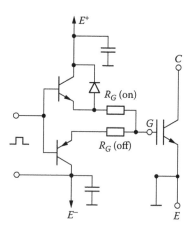

FIGURE 1.17

Model scheme of the IGBT control, providing pulses of opposite polarity on the gate required for reliable blocking and unblocking of the transistor.

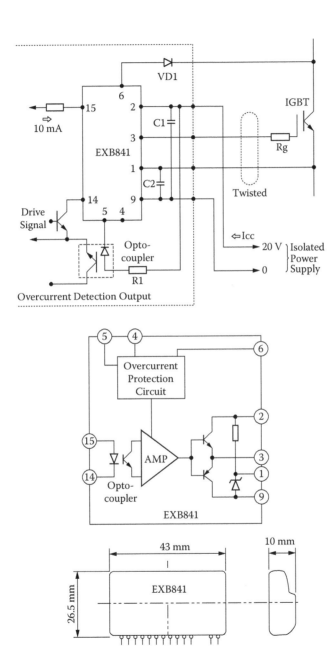

FIGURE 1.18
IGBT-driving hybrid integral circuit EXB841 type (Fuji Electric).

any type of power IGBTs. Apart from forming control signals of the required level and form, such devices often protect the IGBT from short circuits.

1.4 Bipolar Transistor General Modes

As an element of an electric circuit, the transistor is usually used such that one of its electrodes is input, one is output, and the third is the common with respect to the input and output. A transistor is commonly connected to external circuits in a four-terminal configuration, referred to as a "quadrupole." The source of an input signal requiring amplification is connected to the circuit of the input electrode, and the load on which the amplified signal is dissipated is connected to the circuit of the output electrode. Depending on which electrode is common for the input and output circuits, transistor connections fall into three basic circuits, as shown in Figure 1.19.

In the CB (common base) circuit, I_E is the input signal and I_C is the output signal. The current amplification coefficient (also called the "current gain," the ratio of amplifier output current to input current) of a transistor in this configuration is equal to $\alpha = I_C/I_E \approx 1$. A device may have low internal input resistance and high internal output resistance, and, for this reason, a change of the load resistance exerts only a minimal influence on the output current (the functional scheme of this mode relates to current source). CB configurations of transistor connections are not commonly used in practice.

The CE (common emitter) circuit is used most often as an amplification stage. Current gain for this circuit is close to the transistor's gain and is equal to $\beta = I_C/I_B \approx 10$–$200$ and more, depending on the type of transistor used. For direct current, $\beta = h_{FE}$. (h_{FE} is the DC current gain parameter specified by the manufacturer of transistors). This circuit has rather high input resistance (i.e., it does not shunt and weaken the input signal) and low output resistance.

The CC (common collector) circuit is used quite often in cases where it is needed to stage with very high input resistance. The circuit current amplification coefficient is close to that of the DC circuit; however, the main

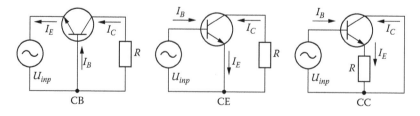

FIGURE 1.19

Transistor basic circuit configurations: (a) common base (CB); (b) common emitter (CE); (c) common collector (CC). I_E—emitter current; I_B—base current; I_C—collector current; R—load.

application of this circuit relates to the functional mode of the voltage—not current amplification—as it is precisely in this mode that one manages to realize the scheme most fully: its very high input resistance. In the voltage amplification mode, the scheme has gain close to one.

Regardless of the transistor used in the circuit, it may function in four main modes determined by the polarity of voltage on the emitter and collector junctions. All possible bias modes are illustrated in Figure 1.20. They are the forward active mode of operation, reverse active mode of operation, saturation mode, and cutoff mode.

As the base current I_B increases or decreases, the operating point moves up or down the load line (Figure 1.21). If I_B increases too much, the operating point moves into the saturation region. In the saturation region, the

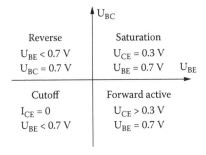

FIGURE 1.20
Possible bias modes of operation of a bipolar junction transistor.

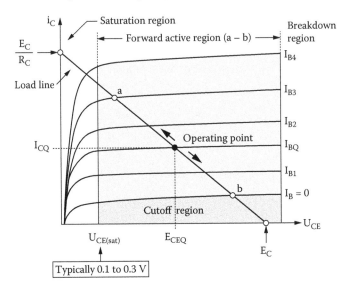

FIGURE 1.21
Output dynamic characteristic of a bipolar junction transistor.

transistor is fully turned ON and the value of collector current I_C is determined by the value of the load resistance R_L. The voltage drop across the transistor V_{CE} is near zero.

In the cutoff region, the transistor is fully turned OFF and the value of the collector current I_C is near zero. Full power supply voltage appears across the transistor. Because there is no current flow through the transistor, there is no voltage drop across the load resistor R_L.

However, bipolar transistors do not have to be restricted to these two extreme mode operations. As described previously, base current "opens a gate" for a limited amount of current through the collector. If this limit for the controlled current is greater than zero but less than the maximum allowed by the power supply and load circuit, the transistor will "throttle" the collector current in a mode somewhere between cutoff and saturation. This mode of operation is called the active mode. A load line is a plot of collector-to-emitter voltage over a range of base currents. The dots (Figure 1.21) marking where the load line intersects the various transistor curves represent the realistic operating conditions for those base currents given.

Let us examine the transistor stage circuit of CE type in detail (Figure 1.22). It begins to get clearer as to what the load line (Figure 1.21) is. This line is built on the series of the static volt-ampere characteristics of transistors (parameters set by the manufacturer) on two cross-points with axes corresponding to the idle mode and short circuit.

Main ratios for this circuit are

$$I_B = \frac{U_B - U_{BE}}{R_B}, \quad I_C = \beta I_B, \quad U_{CE} = E_C - I_C R_C, \quad I_C = \beta I_B, \quad U_{CE} = E_C - I_C R_C$$

For the first of them, $I_C = 0$, $U_{CE} = E_C$ (point on the abscissa axis); for the second, $U_{CE} = 0$, $I_C = E_C/R_C$ (point on the ordinate axis). The intersection points of

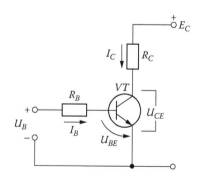

FIGURE 1.22
Transistor stage in CE mode.

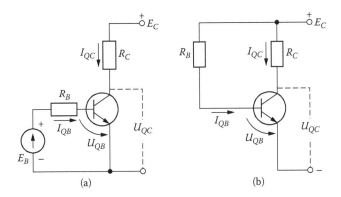

FIGURE 1.23
Circuit diagram of setting of working quiescent point of transistors by fixed current.

the load line with any static characteristics are called the working points corresponding to definite values of output current and output voltage. The transistor functions in the active mode (amplification) when the working point lies within the limits of the a–b interval. The functioning of the transistor stage in the amplification mode is characterized by the so-called "quiescent point" or Q-point (QP). The quiescent operating conditions may be shown on the graph in the form of a single point along the load line (Figure 1.23). For a class A amplifier (widely used in simple automatic devices), the operating point for the quiescent mode (quiescent point) will be in the middle of the load line.

Pragmatically, the working Q-point of a transistor chosen according to its characteristic (Figure 1.21) should be set with the help of the so-called biasing circuit. There are two methods for setting the transistor working point: by fixed current or by fixed voltage. The first one is implemented with the help of two circuit diagrams (Figure 1.23). In the circuit of Figure 1.23(a), the biasing circuit is formed by resistor R_B, which is calculated as follows:

$$R_B = \frac{E_B - U_{QB}}{I_{QB}}$$

where $U_{BE} \approx 0.7$–0.9 V when the base-emitter junction is forward biased; I_{QB} is the quiescent base current from output dynamic characteristic (Figure 1.21)

With only one biasing source, as shown on Figure 1.23(b), the quiescent mode is ensured by the power supply voltage E_C and by resistor R_B:

$$R_B = \frac{E_C - U_{QB}}{I_{QB}}$$

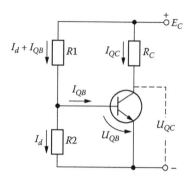

FIGURE 1.24
Circuit diagram of transistor stage with biasing by fixed voltage.

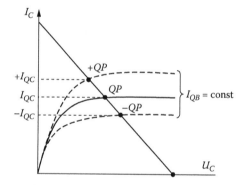

FIGURE 1.25
Changing of transistor quiescent point with displacement direct current I_{QB} temperature influenced.

The method of setting biasing by fixed voltage is shown in Figure 1.24. It is the most widely used method of setting a transistor quiescent point by means of two resistors $R1$ and $R2$.

For this circuit the following ratios are appropriate:

$$R_1 = \frac{E_C - U_{QB}}{I_{QB} + I_d}, \quad R_2 = \frac{U_{QB}}{I_d}, \quad I_d = (2-5)I_{QB}$$

As the environment's temperature changes, the transistor current gain (β) changes (increases as the temperature increases and decreases with a decrease in temperature), and the quiescent point position will change (Figure 1.25).

In this circuit (Figure 1.26), for the resistance of the additional resistor R_E in the emitter circuit, one chooses based on the equation $R_E = (0.1 - 0.2) R_C$ and the capacitance, C, of capacitor C from the equation:

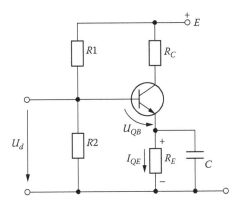

FIGURE 1.26
Amplifying stage with thermostabilization of quiescent point.

$$\frac{1}{\omega C_E} \ll R_E$$

where ω is minimum frequency of the enhanced signal.

Thanks to the biasing capability of this capacitor, in AC applications we obtain the amplification stage with CE and in DC applications the amplification stage with negative feedback.

By the biasing of working points higher than point "a" or lower than point "b," the transistor goes into the saturation or cutoff mode, correspondingly. In the saturation mode, the transistor is fully turned on and the current $I_{SAT} = E_C/R_C$ flows through it; it does not increase more with an increase of the input signal (that is why this mode is called "saturation"). To force the transistor into the saturation mode, one should make its base current not less than $I_B = I_{C(SAT)}/\beta$. In many automation devices, transistors function in the switch mode (i.e., in two ultimate modes: saturation and cutoff). Conditions for safe cutoff or complete saturation of transistor are ensured, as in the case examined previously, by the choice of biasing resistances R_B and R_C.

Equations for voltage in the base circuit are

$$U_B = -I_{C0}R_B + U_{BE} \quad \text{or} \quad U_{BE} = U_B + I_{C0}R_B$$

The safe transistor cutoff is ensured under the condition that $U_{BE} \leq 0$, when $R_B \leq U_B/I_{C0}$. In this case, the value of the base resistance may be calculated as follows:

$$R_B \leq \frac{U_B}{I_{C0\,max}}$$

where I_{C0max} is the maximum value of the collector's reverse current (transistor certified value).

In the circuit diagram of Figure 1.22, the positive input signal of a definite value turns the transistor on (the saturation mode). At this point, currents flowing in the transistor are equal:

$$I_B = \frac{U_B}{R_B}, \quad I_{C(SAT)} = \frac{E_C}{R_C} \leq \beta I_B$$

Thus, one may calculate the value of base resistance for transistor saturation mode:

$$R_B \leq \beta_{min} R_C \frac{U_B}{E_C}$$

1.5 Transistor Devices in Switching Mode

One of the often used operation modes of a transistor is the switching mode; even a single transistor can work as a high speed switch (Figure 1.27).

For switching current from one circuit to another, a two-transistor circuit (Figure 1.28) is used. In this circuit, stable offset voltage is applied to the base of the transistor (T_2) and control voltage to base T_1.

When $u_{inp} = u_{offset}$, the currents and voltages in the arms of the circuit are the same. If the input voltage (u_{inp}) exceeds the offset voltage (u_{offset}), transistor T_2 is gradually blocked and the whole current flows only through transistor T_1 and load resistor R_{C1}, and vice versa. When input voltage decreases below the level of the offset voltage ($u_{inp} < u_{offset}$), transistor T_1 is blocked and T_2 is unblocked, switching the sole current to the circuit of the resistor R_{C2}.

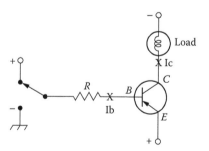

FIGURE 1.27
Electronic switch on a single transistor.

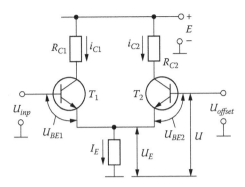

FIGURE 1.28
Transistor switch of two circuits.

As is known, contacts of several electromagnetic relays, connected with each other in a certain way, are widely used in automation systems for carrying out the simplest logical operations with electric signals (Figure 1.29).

For example, the logical operation *AND* is implemented with the help of several contacts connected in series, switched to the load circuit (Figure 1.29a). The signal Y will be the output of this circuit (that is, the bulb will be alight) only if signals on the first input, X1, and on the second input, X2, operate simultaneously (that is, when both contacts are closed). Another simplest logical operation *OR* (Figure 1.29b) is implemented with the help of several contacts connected in parallel. In this circuit, in order to obtain the signal Y on output (that is, for switching the bulb on), input of signal OR on the first input (X1), or on the second input (X2), or on both of the inputs simultaneously is required. Implementation of logical operations with electric circuits is one of the most important functions of relays. Transistor circuits successfully carry out this task. For example, the function NOT can be implemented on any type of single transistor (Figure 1.30).

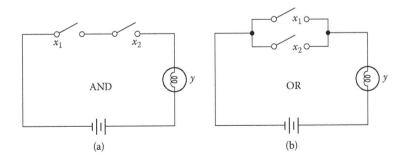

FIGURE 1.29
Implementation of the simplest logical operations, with the help of electromagnetic relay contacts.

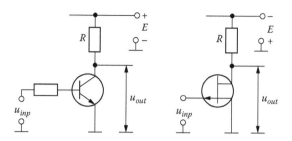

FIGURE 1.30
Logical element NOT implemented on bipolar and field transistors.

In the circuit in Figure 1.30, when an input signal is missing, the transistor is blocked; that is, the whole voltage of the power source E is applied between the emitter and the collector (the drain and the source) of the transistor. Since the output signal is voltage on the collector (the source) of the transistor, that means that if there is no signal at the input, there will be a signal at the output of this circuit. And, vice versa, when the signal is applied at the input, the transistor is unblocked and voltage drops to a very small value (fractions of a volt) and therefore the signal disappears at the output.

The logical element AND-NOT can be implemented by different circuit methods. In the simplest case, this is a circuit from transistors connected in series (Figure 1.31a). When control signals are applied to both inputs X1 and X2 simultaneously, both transistors will be enabled and the voltage drop in the circuit with two transistors connected in series will decrease to a very small value. This means no output signal Y. In the second circuit diagram (Figure 1.31b), even one signal on any input (X1 or X2) is enough for the voltage on output Y to disappear.

Self-contained logical elements are indicated on circuit diagrams as special signs (Table 1.1). A signal strong enough for transition of a logical element

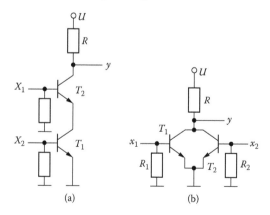

FIGURE 1.31
Transistor logical elements AND-NOT (a) and OR-NOT (b).

TABLE 1.1

Basic Logical Elements

Logical Function	Conventional Symbols	Boolean Identities	Truth Table		
			Inputs		Output
			B	A	Y
And	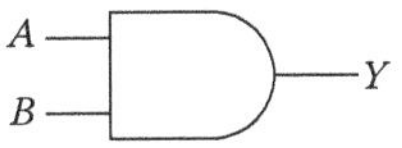	$A \cdot B = Y$	0	0	0
			0	1	0
			1	0	0
			1	1	1
Or	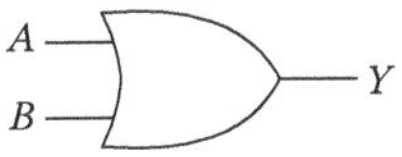	$A + B = Y$	0	0	0
			0	1	1
			1	0	1
			1	1	1
Not	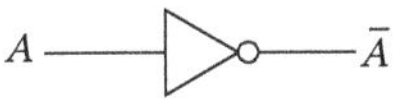	$A + \bar{A}$		0	1
				1	0
And-Not (NAND)	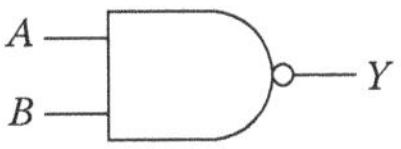	$\overline{A \cdot B} = Y$	0	0	1
			0	1	1
			1	0	1
			1	1	0
Or-Not (NOR)	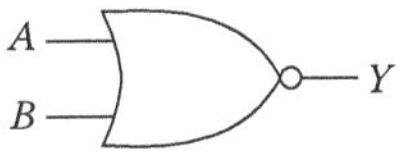	$\overline{A + B} = Y$	0	0	1
			0	1	0
			1	0	0
			1	1	0

Note: According to certain standards' logical elements, which are also indicated as rectangles.

from one state to another is usually marked as "1." No signal (or a very weak signal incapable of affecting the system state) is usually marked as "0."

The same signs are used for indication of the state of the circuit elements: "1" = switched on; "0" = switched off. Such bistable devices (that is, having two stable states) are called triggers. When supply voltage is applied to such a device (Figure 1.32), one of the transistors will be immediately enabled and the other one will remain in a blocked state. The process is avalanche-like and is called regenerative. It is impossible to predict which transistor will be enabled because the circuit is absolutely symmetrical and the likelihood of unblocking both transistors is the same.

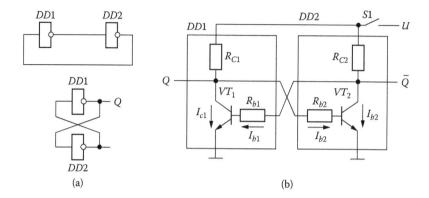

FIGURE 1.32
Bistable relay circuit with two logical elements NOT. (a) chemical symbols; (b) internal circuit diagram.

This state of the device remains stable just the same. Repeated switching ON and OFF of voltage will cause the circuit to pass into this or that stable state. The essential disadvantage of such a trigger is that there is no control circuit, which would enable us to control its state at permanent supply voltage.

In practice the so-called *Schmitt triggers* are often used as electronic circuits with relay characteristics. There are a lot of variants of such triggers, possessing special qualities. In the simplest variant, such a trigger is a symmetrical structure formed by two logical elements connected in a cycle of the type AND-NOT or OR-NOT (Figure 1.33); it is called an *asynchronous RS trigger*.

One of the trigger outlets is named *direct* (any outlet can be named this as the circuit is symmetrical) and is marked by the letter Q, and the other one is called *inverse* and is marked by $\overline{Q}$, to signify that in logical sense the signal at this output is opposite to the signal at the direct output. The trigger state is usually identified with the state of the direct output; that is, the trigger is in the single (that is switched on) state when Q = 1, $\overline{Q}$ = 0, and vice versa.

Trigger state transition has a lot of synonyms—*switching, changeover, overthrow, recording*—and is carried out with the help of control signals applied at the inputs R and S. The input by which the trigger is set up in the single state

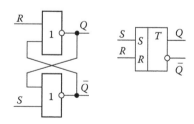

FIGURE 1.33
Asynchronous RS trigger formed by two logical elements NOR.

TABLE 1.2

Combinations of Signals at the Inputs and the RS Trigger Position

Input			Output for logical element type			
			AND-NOT		OR-NOT	
S (set)	R (reset)	Notes	Q	$\overline{Q}$	Q	$\overline{Q}$
0	0	Forbidden mode for AND-NOT	Uncertainty	Without changes		
1	0		1	0	1	0
0	1		0	1	0	1
1	1	Forbidden mode for NOR	Without changes	Uncertainty		

is called the S input (from *set*) and the output by which the trigger turns back to the zero position is the R input (from *reset*). Four combinations of signals are possible at the inputs, each of them corresponding to a certain trigger position (Table 1.2).

As can be seen from Table 1.2, when there are no signals on both of the trigger inputs on the elements AND-NOT (NAND), or when there are signals on both of the trigger inputs on the elements OR-NOT (NOR), the trigger state will be indefinite, which is why such combinations of signals are prohibited for RS triggers.

From the time diagram of the asynchronous RS trigger (Figure 1.34), it can be seen that after transfer of the trigger to the single state, no repeated signals on the triggering input S are capable of changing its state. The return

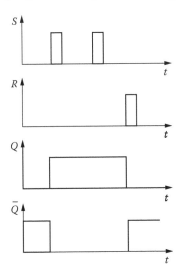

FIGURE 1.34
Time diagram of an asynchronous RS trigger.

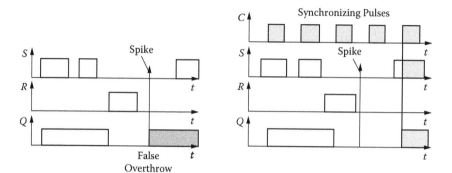

FIGURE 1.35
Time diagrams of operation of an asynchronous trigger (on the left) and synchronous trigger (right) when there is noise.

of the trigger to the initial position is possible only after a signal is applied to its "erasing" R input.

The disadvantage of the asynchronous trigger is its incapacity to distinguish the useful signal of starting from noise occurring in the starting input by chance. Therefore, in practice, so-called *synchronous* or D-triggers, distinguished by an additional so-called *synchronizing input,* are frequently used.

Switching of the synchronous trigger to the single state (ON) is carried only with both signals: starting signal at the S input and also with a simultaneous signal on the synchronizing input. Synchronizing (timing) signals can be applied to the trigger (C input, Figure 1.35) with certain frequencies from an external generator.

A simple amplifier with two transistors, with positive feedback, also has properties of trigger (Figure 1.36).

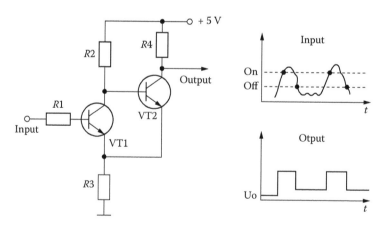

FIGURE 1.36
A simple two-transistor trigger.

In the initial position, when there is no voltage (or voltage is very low) at the input of the circuit, transistor VT1 is closed (locked up). There is voltage on VT1 collector, which opens transistor VT2.

The emitter current of transistor VT2 causes a voltage drop on the resistor R3, which blocks transistor VT1 and holds it in a closed position. If input voltage exceeds the voltage in the emitter on the VT1 transistor, it will be opened and will become saturated with very small collector–emitter junction resistance.

As a result, the potentials of the base and the emitter of transistor VT2 will be equal. Transistor VT2 will be blocked. At the output there will be voltage equal to the supply voltage. When input voltage decreases, transistor VT1 leaves the saturation mode, and an avalanche-like process occurs. Emitter current of transistor VT2, causing blocking voltage on resistor R3, accelerates closing of the transistor VT1. As a result, the trigger returns to its initial position.

1.6 Thyristors

The history of development of another remarkable semiconductor device with relay characteristics begins with the conception of a "collector with a trap," formulated at the beginning the 1950s by William Shockley, who is familiar to us from his research on p–n junctions. Following Shockley, J. Ebers invented the two-transistor analogy (interbounded n–p–n and p–n–p transistors) of a p–n–p–n switch, which became the model of such a device (Figure 1.37).

The working element of this new semiconductor device with relay characteristics was a four-layer silicon crystal with alternating p- and n-layers

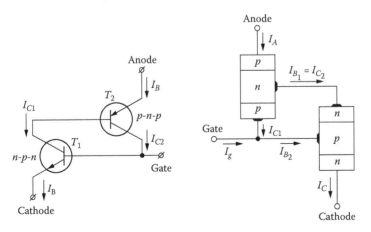

FIGURE 1.37
Two-transistor model of a thyristor.

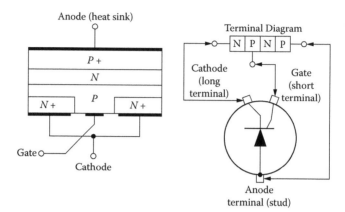

FIGURE 1.38
Structure and symbolic notation of a solid-state thyratron—"thyristor."

(Figure 1.38). Such a structure is made by diffusion into the original mono-crystal of n_1-silicon (which is a disk 20–45 mm in diameter and 0.4–0.8 mm thick, or more for high voltage devices) admixture atoms of aluminum and boron from the direction of its two bases to a depth of about 50–80 μm. Injected admixtures form p_1 and p_2 layers in the structure.

The fourth (thinner) layer n_2 (its thickness is about 10–15 μm) is formed by further diffusion of atoms of phosphorus to the layer p_2. The upper layer, p_1, is used as an anode in the thyristor and the lower layer, p_2, as a cathode.

The power circuit is connected to the main electrodes of the thyristor: the anode and the cathode. The positive terminal of the control circuit is con-nected through the external electrode to layer p_2 and the negative one to the cathode terminal.

The volt-ampere characteristic (VAC) of a device with such a structure (Figure 1.39) much resembles the VAC of a diode by form. As in a diode, the VAC of a thyristor has forward and reverse areas. Like a diode, the thyristor is blocked when reverse voltage is applied to it (minus on the anode, plus on the cathode) and when the maximum permissible level of voltage (U_{Rmax}) is exceeded, there is a breakdown, causing strong current and irreversible destruction of the structure of the device.

The forward area of the VAC of the thyristor does not remain permanent, as does that of a diode but can change, being affected by the current of the control electrode, called the *gate*. When there is no current in the circuit of this electrode, the thyristor remains blocked not only in reverse but also in the forward direction; that is, it does not conduct current at all (except small leakage current, of course). When the voltage applied in the forward direction between the anode and the cathode is increased to a certain value, the thyris-tor is quickly (stepwise) enabled and only a small voltage drop (fractions of a volt), caused by irregularity of the crystal structure, remains on it. If low current is applied to the circuit of the gate, the thyristor will be switched ON

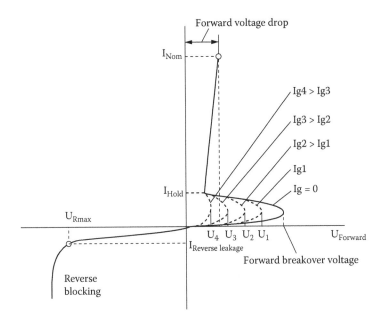

FIGURE 1.39
Volt-ampere characteristic (VAC) of a thyristor.

to much lower voltage between the anode and the cathode. The higher the current is, the lower is the voltage required for unblocking of the thyristor. At a certain current value (from a few milliamperes for low-power thyristors up to hundreds of milliamperes for power ones), the forward branch of the VAC is almost fully rectified and becomes similar to the VAC of a diode. In this mode (that is, when control current constantly flows in the gate circuit), the behavior of the thyristor is similar to that of a diode that is fully enabled in the forward direction and fully blocked in the reverse direction. However, it is senseless to use thyristors in this mode; there are simpler and cheaper diodes for this purpose.

In fact, thyristors are used in modes when the working voltage applied between the anode and the cathode does not exceed 50%–70% of the voltage, causing spontaneous switching ON of the thyristor (when there is no control signal, the thyristor always remains blocked). Control current is applied to the gate circuit only when the thyristor should be unblocked and of such a value that would enable reliable unblocking. In this mode, the thyristor functions as a very high speed relay (unblocking time is a few or tens of microseconds).

Perhaps many have heard that thyristors are used as basic elements for smooth current and voltage adjusting, but if a thyristor is only an electric relay having two stable states like any other relay—a switched ON state and a switched OFF one—how can a thyristor smoothly adjust voltage? The point is that if nonconstant alternating sinusoidal voltage is applied, it is possible to adjust the unblocking moment of the thyristor by changing the moment

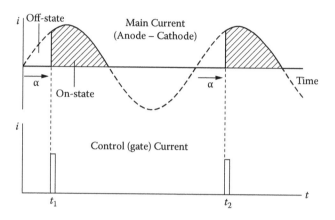

FIGURE 1.40
Principle of operation of a thyristor regulator.

of applying a pulse of control current on the gate with regard to the phase of the applied forward sinusoidal voltage. That is, it is as though a part of the sinusoidal current flowing to the load were cut off (Figure 1.40). The moment of applying a pulse of unblocking control current (such pulses are also called "igniting" by analogy with the control pulses of the thyratron) is usually characterized by the angle α.

Taking into account that average current value in the load is defined as an integral (that is, the area of the rest of the part of the sinusoid) the principle of operation of a thyristor regulator becomes clear. After unblocking, the thyristor remains in the opened state, even after completion of the control current pulse. It can be switched OFF only by reducing forward current in the anode–cathode circuit to the value less than *hold current* value.

In AC circuits, the condition for thyristor blocking is created automatically when the sinusoid crosses the zero value. To unblock the thyristor in the second half-wave of the voltage, it is necessary to apply a short control pulse through the gate of the thyristor.

To control both half-waves of alternating current, two thyristors connected antiparallel are used. Then, one of them works on the positive half-wave and the other on the negative one. At present such devices are produced for currents of a few milliamperes to a few thousand amperes, and for blocking voltages up to a few thousands volts (Figure 1.41).

1.7 Optocouplers

Optocouplers (optical couplers) are designed to isolate electrical output from input for complete elimination of noise. All digital inputs of DPRs

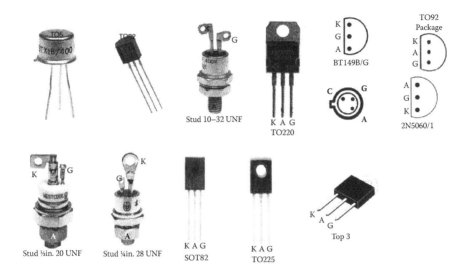

FIGURE 1.41
Modern low-power and power thyristors.

are connected to the internal circuit of the DPR and to the CPU through optocouplers.

The blocked $n–p$ junction in semiconductor devices (diodes, transistors, thyristors) may begin to allow electric current to pass under the effect of energy of photons (light). When the $n–p$ junction is illuminated, additional vapors of charge carriers—electrons and holes causing electric current in the junction—are generated within it. The higher the intensity of the luminous flux on the $n–p$ junction is, the stronger the current is. Optoelectronic relays (Figure 1.42) comprise a light-emitting element that is usually made on the basis of a special diode (light emission diode [LED]), an $n–p$ junction emitted by photons when current passes through it, and a receiver of the luminous flux (a photodiode, phototransistor, photothyristor).

Optocouplers typically come in a small six-pin or eight-pin IC package, but are essentially a combination of two distinct devices: an optical transmitter, typically a gallium arsenide LED, and an optical receiver such as a phototransistor or light-triggered DIAC (Figure 1.42). The two are separated by a transparent barrier that blocks any electrical current flow between the two, but does allow the passage of light. The basic idea is shown in Figure 1.43, along with the usual circuit symbol for an optocoupler. Usually, the electrical connections to the LED section are brought out to the pins on one side of the package and those for the phototransistor or DIAC to the other side, to separate them physically as much as possible. This usually allows optocouplers to withstand voltages of anywhere from 500 to 7500 V between input and output. Optocouplers are essentially digital or switching devices, so they are best for transferring either on–off control signals or digital data.

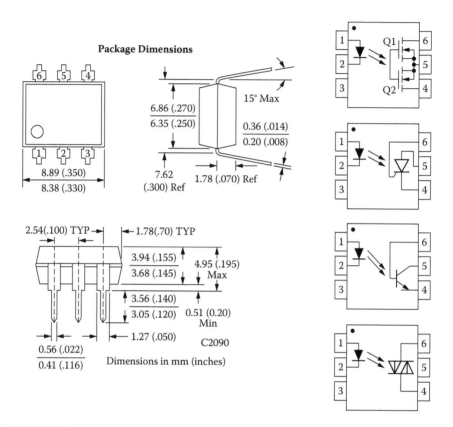

FIGURE 1.42
Optoelectronic relays designed in a standard DIP case.

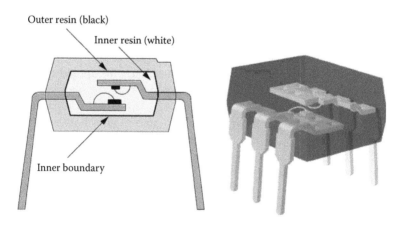

FIGURE 1.43
Internal design of the optocoupler.

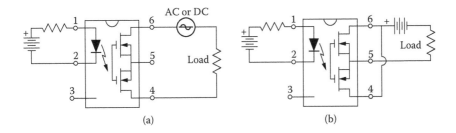

FIGURE 1.44
Connection of bidirectional switched photo-MOS optocoupler.

Usage of the two series connected photo-MOS transistors ("A" connection on Figure 1.44) as output elements allows the optoelectronic relay to switch either AC or DC loads with nominal output current rating. Connection "B" with the polarity and pin configuration, as indicated in the schematic, allows the relay to switch DC load only, but current capability increases by a factor two.

There is a great diversity of circuits and constructions of power optocouplers (optoelectronic relays), including those containing built-in invertors or amplifiers (Figure 1.45). A similar principle serves as a basis not only for miniature devices in chip cases, but also for practically all power semiconductor relays and contactors. It should be noted that the external designs of not only miniature optocouplers in chip cases, but also of more powerful devices of various firms, are very much alike (Figure 1.46).

Such relays are usually constructed according to a similar scheme (Figure 1.47), with only some slight variations. As a rule they comprise an RC circuit (the so-called "snubber") and a varistor protection outlet, protecting the thyristors from overvoltages. They often contain a special unit (a zero voltage detector) controlling the moment when the voltage sinusoid passes

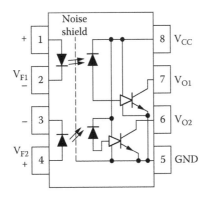

FIGURE 1.45
Twin optoelectronic relay with built-in amplifiers of power.

FIGURE 1.46
Modern semiconductor optoelectronic relays for currents of 3–5 A, produced by different companies.

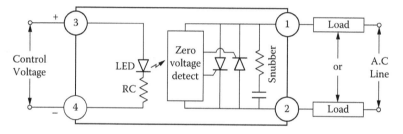

FIGURE 1.47
Standard scheme of a power single-phase optoelectronic AC relay.

through the zero value and allowing it to enable (and sometimes to disable) the thyristor at the zero value of voltage (so-called "synchronous switching").

Optocoupler characterizes by current transfer ratio (CTR), which is defined as the ratio of the output current (I_C) to the input current (I_F).

The brightness of the LED slowly decreases in an exponential fashion as a function of forward current (I_F) and time. A 50% degradation is considered to be the failure point; therefore, an optocoupler's lifetime is defined as the time at which the CTR has fallen to 50% of its original value. According to Slama et al. [1], at LED current equal to 0.5 mA and ambient temperature 25°C, the optocoupler's lifetime will be equal to 2.10^5 hours (~22 years). In some critical applications, the malfunction appears already at CTR decreasing up to 70%–80% (Figure 1.48)—that is, upon 10–11 years during normal electronic equipment lifetime.

1.8 Electromagnetic Relays

An electromagnetic neutral relay is the simplest, most ancient, and most widespread type of relay. What are its basic elements? As a rule, most people

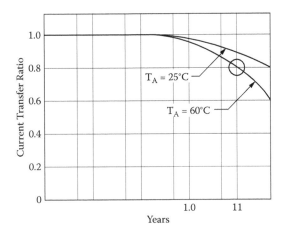

FIGURE 1.48
CTR degradation over time at 25°C and 60°C ambient temperature.

asked this question would probably name the following: a winding, a magnetic core, an armature, a spring, and contacts.

This all is true, of course, but if one begins to analyze how a relay works, the thought might occur that there is something missing. What is the purpose of a magnetic system? Apparently it is used to transform input electric current to the mechanical power needed for contact closure. And what does a contact system do? It transforms the imparted mechanical power back to an electric signal!

Is something wrong here?

Everything will become more obvious if the list of basic components of a relay includes one more element, which is not so obvious from the point of view of the construction of a relay—for example, a coil or contacts. Very often it is not just one element but several small parts that escape our attention. Such parts are often omitted on diagrams illustrating the principle of relay operation (Figure 1.49).

I am referring to an insulation system providing galvanic isolation of the input circuit (winding) from output one (contacts). If we take such an insulation system into account, it becomes clear that an input signal at the relay input and an output signal at the relay output are not the same. They are two different signals that are completely insulated from each other electrically.

Note Figure 2.1, which is often used to illustrate principles of relay operation. If one uses this figure as the only guide while constructing a relay, the relay will not operate properly since its input circuit (the winding) is not electrically insulated from the output circuit (the contacts).

In simple constructions used for work at low voltage, insulating bobbins with winding (shown on the Figure 1.50) provide basic insulation (apart from an insulator). The magnetic system of a typical low voltage electromagnetic relay comprises first of all a control winding (6) made in the form of a

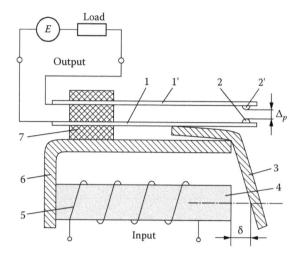

FIGURE 1.49
Construction of a simple electromagnetic relay: 1—springs, 2—contacts, 3—armature, 4—core, 5—winding, 6—magnetic core, 7—insulator.

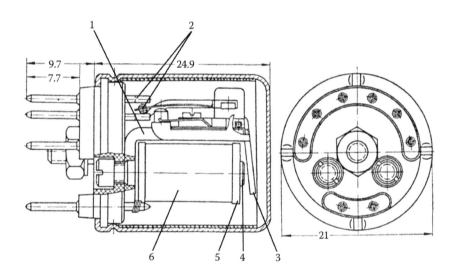

FIGURE 1.50
Real construction of low-voltage miniature relay: 1—magnetic core; 2—contacts; 3—armature; 4—pole of core; 5—insulating bobbin; 6—coil.

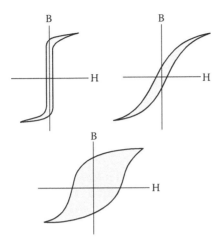

FIGURE 1.51
Hysteresis loops for soft and hard magnetic materials. B—flux density; H—magnetic field strength.

coil with insulated wire, a magnetic core (1), and a movable armature (3) (see Figure 1.50).

Elements of the magnetic circuit of the relay are usually made of soft magnetic steel—a type of steel that has a small hysteresis loop (Figure 1.51). *Hysteresis* can be translated as a "lag." A hysteresis loop is formed from a magnetization curve and a demagnetization curve. These curves are not superimposed on each other because the same magnetic field strength is not enough to demagnetize the magnetized material. An additional magnetic field is required for demagnetization of the material to the initial (nonmagnetized) state. This happens because the so-called residual induction still remains within the previously magnetized model, even after removal of the external magnetic field.

Such a phenomenon takes place because magnetic domains (crystal structures of ferromagnetic material) expanding along the external magnetic field lose their alignment when the external magnetic field is removed. The demagnetizing field necessary for a complete demagnetization of the previously magnetized model is called the *coercive force*. The coercive force of soft magnetic materials is small, which is why their hysteresis loop is also small. That means that when the external magnetic field is removed (de-energizing the winding of the relay), the magnetic core and armature do not remain magnetized, but return to their initial stage. Apparently this lack of retentivity is a very important requirement for materials used in magnetic circuits of typical neutral relays. Otherwise, relay characteristics would not be stable and the armature might seal.

Electric steel used for the production of relays is composed of this very soft magnetic material. This is steel with a lean temper (and other admixtures such as sulfur, phosphor, oxygen, and nitrogen) and rich silicon (0,5 – 5). Apart from enhancing the qualities of the steel, the silicon makes it more stable and increases electrical resistance, which considerably weakens eddy currents (see later discussion). The hardness and the fragility of the steel are mostly dependent on the silicon content. With a content of 4%–5% of silicon, steel usually can withstand no more than one to two folds of 90°.

For a long time, only the hot rolling method was used for the production of such steel; then, in 1935, Goss discovered the superior magnetic properties of cold-rolled electric steel (but only along the direction of rolling). That gave such steel a magnetic texture and made it anisotropic. The utilization of anisotropic cold-rolled steel requires such construction of the magnetic core, to enable magnetic flux to pass only in the direction of rolling. Machining of parts of a magnetic core entails high internal stress and, consequently, higher coercive force. That is why after processes of forming, whetting, and milling, parts must be annealed at 800°C–900°C, with further gradual reduction of temperature to 200°C–300°C.

Sometimes Permalloy is used for magnetic cores of highly sensitive relays. It is a ferroalloy with nickel (45%–78%) alloyed with molybdenum, chrome, copper, and other elements. Permalloy has better magnetic properties in weak magnetic fields than electric steel; however, it cannot be used for work at great magnetic fluxes since its saturation induction is only half as much as that of electric steel.

Magnetic cores for big AC relays are made of sheet electric steel, which is 0.35–0.5 mm thick. Several types of the magnetic systems are used in modern constructions of relays.

The clapper-type (attracted-armature) magnetic systems is the most ancient type of magnetic systems. Its construction was already described in Edison's patents. It was used first in telephone relays and later in industrial and compact covered relays. Today these types of magnetic system are widely used in constructions of middle-sized and small relays, with a plastic rectangular cover that is often transparent. They are mostly designed for work in systems of industrial automation and the power-generation industry (Figure 1.52). The disadvantage of relays with this type of magnetic system is sensitivity to external mechanical effects.

Further research in this direction led to the invention of a new J-type magnetic system. In this system, a balanced turning armature was widely used in miniature hermetic relays, and the rotation axis of the armature goes through its center of mass. As a result, the relay becomes resistant to external mechanical impact and can sustain linear accelerations of 100–500 g, reiterated shocks at an acceleration of 75–150 g, and separate shocks at an acceleration rate of up to 1000 g. There are a lot of variants of this J-type magnetic system (Figure 1.53).

Not only most modern miniature relays in plastic shells, but also hermetic relays in metal shells (Figure 1.54), which have been produced for ages, have similar magnetic systems. Miniature and microminiature relays of this type

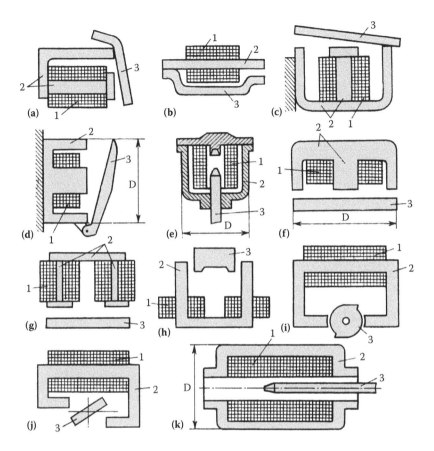

FIGURE 1.52
Types of magnetic systems of modern electromagnetic relays: 1—control coil; 2—magnetic core; 3—armature. (a–d)—clapper type (attracted-armature) systems; (f, g)—direct motion magnetic systems; (e, h, k)—systems with a retractable armature (solenoid type); (i, j)—systems with a balanced turning armature.

are really very small in size: For example, the RES49 type relay is 10.45 × 5.3 × 23.2 mm with a weight of 3.5 g or the RES80 type relay is 10.4 × 10.8 × 5.3 mm with a weight of 2 g.

Another kind of the electromagnetic relay widely used in MPD is a latching relay. This is one that picks up under the effect of a single current pulse in the winding and remains in this state when the pulse stops affecting it—that is, when it is locked. Therefore, this relay plays the role of a memory circuit. Moreover, a latching relay helps reduce power dissipation in the application circuit because the coil does not need to be energized all the time (Figure 1.55).

As illustrated in Figure 1.55(b), the contacts of a latching relay remain in the operating state even after an input to the coil (set coil) has been removed. Therefore, this relay plays the role of a memory circuit. The double coil latch

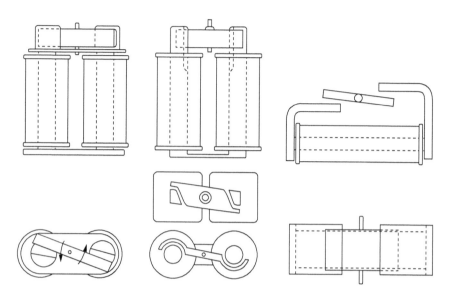

FIGURE 1.53
Some variants of construction of the J-type magnetic system.

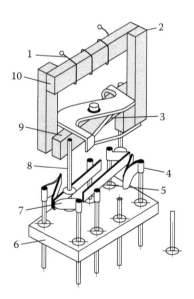

FIGURE 1.54
One of the most popular variants of construction of the turning-type magnetic system, widely used in miniature relays: 1—winding; 2—pole of the magnetic core; 3—restoring spring; 4—fixed normally open contact; 5—moving contact; 6—heel piece of relay; 7—fixed normally closed contact; 8—pusher with a small insulating ball, weighted at the end; 9—turning armature; 10—core.

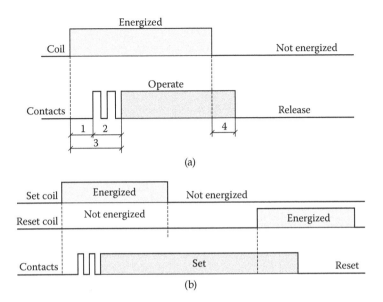

FIGURE 1.55
(a) Time chart of nonlatch relay: 1—current-rise time; 2—bounce time; 3—full operate time; 4—release time. (b) Time chart of double coil latch relay.

type relay has two separate coils, each of which operates (sets) and releases (resets) the contacts.

In latch relays, two types of latching elements are usually used: magnetic and mechanical. As with standard electromagnetic relays, latching relays are produced for all voltage and switched power classes: from miniature relays for electronics, with contact systems and cases typical of standard relays of the same class according to switched power, up to high voltage relays and high current contactors.

Magnetic systems of most miniature latching relays are constructed on the base of permanent magnets (Figures 1.56 and 1.57). The smallest latching relays in the world in standard metal cases of low power transistors are produced by the American company Teledyne Relays (Figure 1.58).

Another popular type of latching relay with magnetic latching is the so-called *remanence* type. This relay consists of a coil and an armature made of a special ferromagnetic material on a nickel base, with admixtures of aluminum, titanium, and niobium. It is capable of becoming magnetized quickly under the effect of a single current pulse in the coil and of remaining in its magnetized state when the pulse stops affecting it.

This type of relay contains a coil with one or two windings wound around it. In the first case, magnetization and demagnetization of the material of the core is carried out by current pulses of opposite polarities and, in the second case, by two different windings on the same bobbin, one of which magnetizes the core while the other one, which demagnetizes it, is a disabling one.

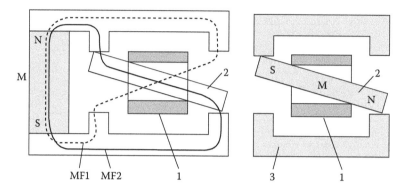

FIGURE 1.56
Popular types of the magnetic system of cheap miniature latching relays, in plastic cases produced by many companies. M—permanent magnet; MF1—magnetic flux in first position; MF2—magnetic flux in second position; 1—coil; 2—rotating armature; 3—yoke.

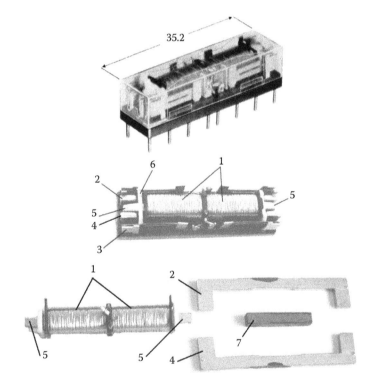

FIGURE 1.57
Construction of a miniature latch relay with a magnetic latching of the DS4 type, produced by Euro-Matsushita: 1—set and reset coils; 2 and 4—plates of the magnetic core; 3—contacts; 5—ferromagnetic pole lugs; 6—plastic pushers put on the pole lugs; 7—permanent magnet placed in the center of the coil.

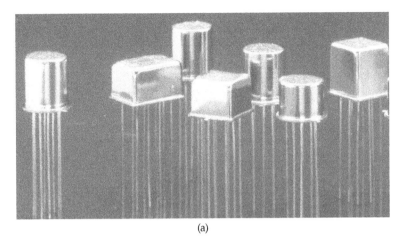

(a)

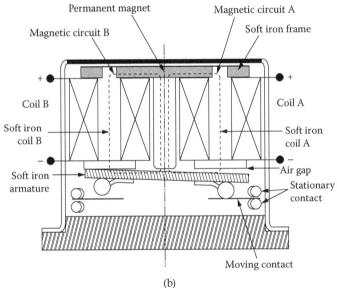

(b)

FIGURE 1.58
(a) The smallest latching relays in the world, produced by the American company Teledyne Relays. External design of relays in Centigrid° and TO-5 cases. (b) Construction of miniature latching relays produced by the Teledyne Relays (United States).

The advantage of this type of relay is that it does not require any special construction. One has only to make a core in the already existing construction of a standard relay of any type from remanent material, and the latching relay is complete!

Miniature single-coil latching relays for operating voltage of 3–5 V are found in many applications, including MPD. In these relays, coil current must flow in both directions through a single coil (Figure 1.59). Current

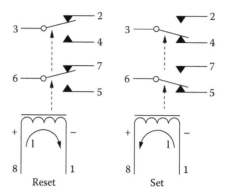

FIGURE 1.59
Principle of operation of a miniature single coil latching relay. The AROMAT AGN210A4HZ relay, for example, has such a principle.

flowing from pin 8 to pin 1 causes the relay to latch in its reset position, and current flowing from pin 1 to pin 8 latches the relay in its set position.

A simple integral circuit of the MAX4820/4821 type (+3.3 V/+5 V, eight-channel, cascadable relay drivers with serial/parallel interface), produced by Maxim Integrated Products Ltd., drives up to four such single-coil (and also four ordinary dual-coil) latching relays. It includes a parallel-interface relay driver (UI) with open-drain outputs (Figure 1.60) and inductive-kick-back protection. Latch any of the four relays to their set or reset positions by turning on the corresponding output (OUTX). That output is selected by asserting its digital address on pins A2 to A0 while CS is high. Activate the output by toggling CS. Both devices feature separate set and reset functions that allow the user to turn ON or turn OFF all outputs simultaneously with a single control line. Built-in hysteresis (Schmidt trigger) on all digital inputs allows this device to be used with slowly rising and falling signals such as those from optocouplers or RC power-up initialization circuits.

The MAX4820 features a digital output (DOUT) that provides a simple way to daisy chain multiple devices. This feature allows the user to drive large banks of relays using only a single serial interface (Figure 1.61).

Electromechanical relays used in MPD are amplifier-driven relays (i.e., relays controlled by electronic switches). Very often, single bipolar (Figure 1.62) or field-controlled (Figure 1.63) transistors are used as amplifiers.

Diodes switched parallel to the winding of the relay are necessary for preventing damage to transistors by overvoltage pulses occurring on the winding of the relay at the moment of blocking of the transistors (transient suppression).

Especially for electromagnetic relay control, a set of amplifiers on the basis of Darlington's transistor is produced in a standard case of an integrated circuit (Figure 1.64). The eight *n–p–n* Darlington connected transistors in this family of arrays are ideally suited for interfacing between low logic

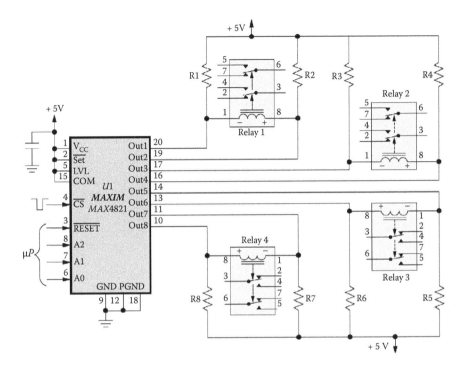

FIGURE 1.60
The MAX4821 integral circuit easily drives four single-coil latching relays.

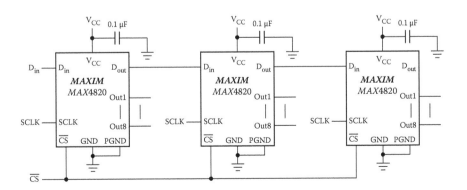

FIGURE 1.61
Daisy-chain configuration for MAX4820.

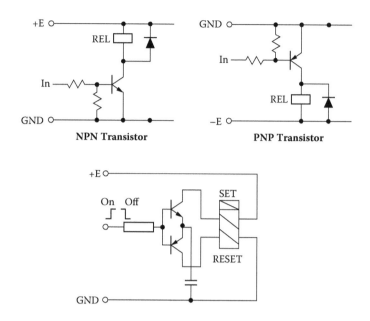

FIGURE 1.62
Amplifier-driven relays on bipolar transistors.

level digital circuitry (such as transistor–transistor logic [TTL], complementary metal oxide semiconductor [CMOS], or positive [P]MOS/negative [N] MOS) and the higher current/voltage relays for a broad range of computer, industrial, and consumer applications. All these devices feature open-collector outputs and free-wheeling clamp diodes for transient suppression. The ULN2803 is designed to be compatible with standard TTL families, while the ULN2804 is optimized for 6 to 15 V high-level CMOS or PMOS.

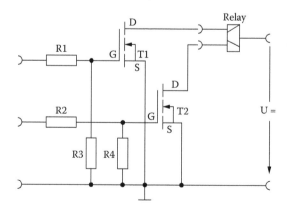

FIGURE 1.63
Amplifier-driven latching relay on FET transistors.

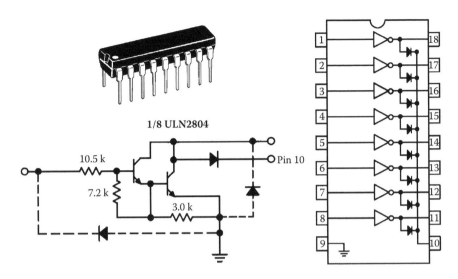

FIGURE 1.64
UNL2804 type chip (Motorola).

Amplifier-driven relays are produced by some firms in the form of independent, fully discrete devices with the electronic amplifier built in directly to the case of the electromagnetic relay. Even the smallest electromagnetic relays in the world, for military and industrial applications, produced by Teledyne, are placed in transistor cases (TO-5 type packages, see previous discussion) and are produced with built-in amplifiers (Figure 1.65).

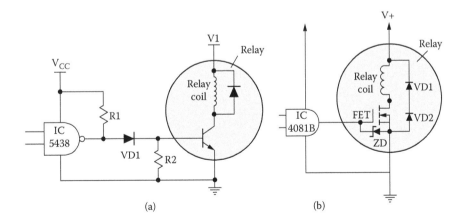

FIGURE 1.65
Amplifier-driven super miniature relays on bipolar (a) and FET (b) transistors, produced by Teledyne Relays Co.: (a)—circuit diagram for relay series: 411, 412, 431, 432; (b)—circuit diagram for relay series: 116C and 136C.

Reference

1. Slama, B. H. J., H. Helali, A. Lahyani, K. Louati, P. Venet, and G. Rojat. 2008. Study and modeling of optocouplers ageing. *Journal of Automation & Systems Engineering*, 2(3). (http://jase.esrgroups.org/edition-2008.php)

2

Secondary Power Supplies

2.1 Introduction

There is no exact definition for the term "secondary power supplies" and different authors give different definitions of this term. We shall understand this term as the power supplies intended for transformation of a main voltage of the electrical network (DC or AC) into voltages of other classes, intended for the direct power supplies of electronic devices. Secondary power supplies can be external (in the form of separate modules) or built in—that is, integrated in electronic devices and being their integral part. Such a secondary power supply is the principal component of any electronic device, including relay protection devices and specially digital protective relays (DPRs) upon which the reliability of the device's working capacity depends. Today, a DPR is used exclusively for switch-mode power supplies.

2.2 Comparative Characteristics of Linear and Switch-Mode Power Supplies

In the 1960s the first switch-mode power supply (SMPS) was brought out And, since then, it has been intensively developed until today the SMPS has almost completely eclipsed the older linear power supply (LPS) from all areas of technology. What is the difference between these two types of the secondary power supplies? Is an SMPS superior to an LPS?

Widely applied everywhere in technical equipment during many decades, the LPSs are rather simple and even primitive devices (Figure 2.1), consisting of only a few elements: a voltage transformer, the rectifier, a filter based on a capacitor, and a semiconductor stabilizer (the Zener diode with the powerful transistor or a single-power solid-state element with an analogous function).

Unlike LPSs, SMPSs are much more complex devices working at high frequency and consisting of hundreds of active and passive elements (Figure 2.2).

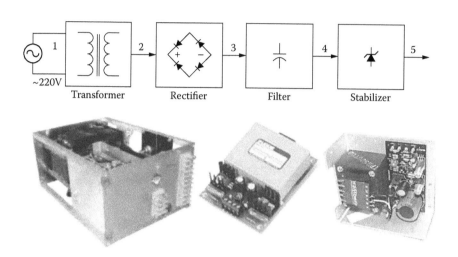

FIGURE 2.1
Structure and appearance of linear power supplies.

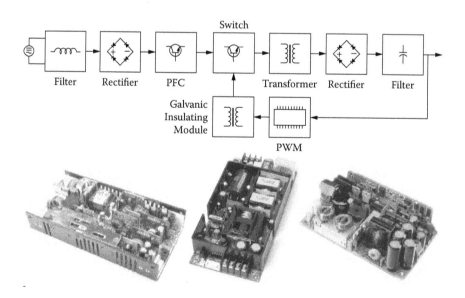

FIGURE 2.2
Structure and appearance of switching mode power supplies.

What are the basic differences between these two types of power supplies? In the LPS the input alternating voltage is transformed to a necessary level (or levels, in the case of multiple secondary windings in the transformer) by means of the transformer and then rectified by a diode bridge, filtered by means of the electrolytic capacitor, and stabilized by a nonlinear electronic element. The voltage applied to the stabilizing element must be greater than the nominal output voltage of the power supply, and its excess is dissipated in the form of heat on this stabilizing element. (This sometimes demands the use of a heat sink.)

The presence of some excess of a voltage on the stabilizing element enables carrying out stabilization of an output voltage with decreased or increased input voltage due to a change of the share of the energy dissipation on the stabilizing element. For this reason, the coefficient of efficiency of such power supplies is always much below one.

In the SMPS the input alternating voltage is first of all rectified by a diode bridge (or simply passed through the diodes of this bridge without change in the case of feeding a secondary power supply from a DC network). Then it is smoothed out and acts on the switching element (usually based on a metal oxide semiconductor field effect transistor [MOSFET]) by means of which the constant voltage is "cut" into narrow strips (switching frequency is from 70 up to 700 kHz for high-power supplies and from 1 to 3 MHz for low-power supplies). The rectangular high-frequency pulses that are generated are applied to the transformer, which outputs voltage matching the demanded level of a voltage that then is rectified and smoothed. The stabilization of the level of the output voltage at changes of the level of the input voltage is carried out by means of a feedback circuit consisting of a special driver that provides a pulse-width modulation (PWM) control signal of the switching element through a galvanic decoupling unit. (It is usual to include an additional isolation transformer.) This driver is small, but contains a complex integral circuit to change the width of control pulses according to power supply output voltage level in order to compensate for deviations.

Low cost power supplies have such structures. Better and more expensive SMPS devices contain at least two additional units: the input high-frequency filter and the power factor corrector (PFC). The first unit is necessary for protection of the network—that is, all other consumers connected to the same network as SMPS from the high-frequency harmonics oscillated into the network by the SMPS. The second unit is used to increase a power factor of the power supply.

The problem of the correction of a power factor (PF) originates due to the presence of the rectifier bridge with the smoothing capacitor on SMPS input. In such a circuit, the capacitor consumes a current, by pulses, from a network only during those moments of time when an instantaneous value of the input sinusoidal voltage becomes more than the DC voltage on the capacitor (which depends on that discarded from the load). During the rest of the time when the voltage on the capacitor is more than the instantaneous input value,

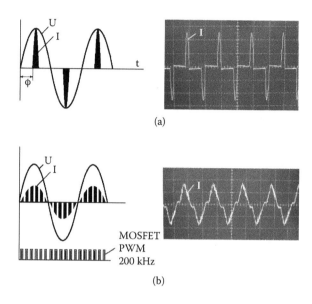

(a)

(b)

FIGURE 2.3

The form of a current and shift of phases between a voltage and a current consumed SMPS: (a) without PFC and (b) with PFC.

diodes of a rectifier bridge appear locked by a reverse voltage applied from the capacitor, and consumption of a current is absent. As a result, the current consumed by the SMPS appears essentially out of voltage phase (Figure 2.3).

When a great number of SMPS devices are connected to an AC network, the combined decrease of the PF in the network becomes appreciable. (Typical PF value for a single SMPS without correction is 0.65.) In this connection, active PF correction is employed by means of the so-called power factor corrector in the SMPS.

PFC is an independent voltage converter, the so-called "boost converter" (BC), supplied with the special control circuit (Figure 2.4). Basic elements of

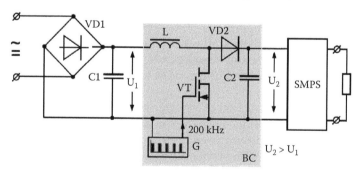

FIGURE 2.4

Connections diagram of the buster converter (BC) with the SMPS.

the BC are choke, L; diode, VD2; capacitor, C2; and fast switching element, VT (based on a MOSFET). The functioning of this device is based on the production of a high-voltage pulse with a reversed polarity on the inductance (L) at the breaking of a current in its circuit.

Transistor VT switches the current in the inductance, L, on and off at a high frequency (it is usually 200 kHz), and the high-voltage pulses formed during the switching process charge the capacitor, C2, through diode VD2 from which the load (in our case, actually, the SMPS) feeds. Thus, a voltage on the capacitor C2 always exceeds the input BC voltage. Owing to this property, the BC is widely used in electronic devices as the voltage level converter: from the standard voltaic cell voltage level of 1.2–1.5 V to the other standard voltage level of 5 V, which is necessary for integral microcircuit control. In our case, the capacitor C2 is charged up to a voltage of 385–400 V.

Owing to that, capacitor C1 has a very small capacity (it is, per se, only a high-frequency filter) and the control circuit of the switching element with PWM constantly traces a phase of the input alternating voltage and provides a matching binding of the control pulses (that is, pulses of a current carried through the switching element) to a phase of an applied voltage. It is possible to eliminate completely the phase shift between the current and the voltage consumed by capacitor C2 (Figure 2.4). Additionally, the same control circuit provides rigid stabilization of a level of voltage charge on the capacitor C2. Despite the small dimensions of the PFC control microchip, it has a complex internal structure (Figure 2.5), and the entire PFC unit is considerably complex and occupies an appreciable area on the SMPS printed circuit board, because of the number of additional passive elements (Figure 2.6).

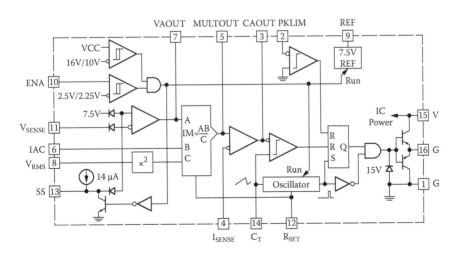

FIGURE 2.5
A microcircuit L6561 type (STMicroelectronics) for buster converter controls–power factor corrector.

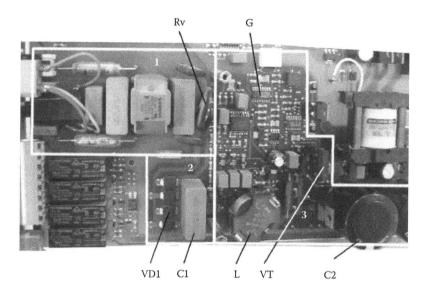

FIGURE 2.6

A fragment of printed circuit board with PFC and BC: 1—input filter; 2—input rectifier bridge with the filtering capacitor; 3—PFC; RV—varistor; G—microcircuit for buster converter controls.

The question arises: Why have such complex devices as SMPSs expelled such simple and well-proven LPS devices from the market?

The basic SMPS advantages over the LPS that are usually specified in the technical literature are the following:

1. Significant decrease in size and weight due to the smaller main transformer (The high-frequency transformer has considerably smaller dimensions and weight in comparison with the transformer of a commercial frequency of the same power.)
2. The very wide range of a working input voltage
3. Considerably higher coefficient of efficiency (up to 90%–95% against 40%–70% for LPS)

In addition to these reasons, we would also add one more important advantage: the possibility of working in a network of both AC and DC voltages.

At first glance (Figure 2.7), the differences between two devices, LPS (on the left) and SMPS (on the right), equal in power and properties are readily apparent: The LPS is much simpler, but contains a much larger and heavier transformer (T).

The flat module SMPS (Figure 2.7, on the right) is the universal power supply of microprocessor-based protective relays of such series as SPAC, SPAD, SPAU, etc., which is moved in the relay case. Naturally, to use the relay design

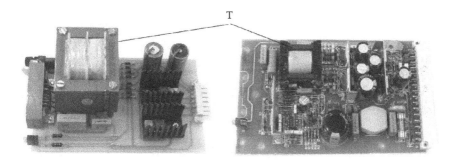

FIGURE 2.7
The linear (at the left) and switch-mode (on the right) power supplies with identical characteristics; T—the transformer.

of LPS with the large transformer is inconvenient. But, what prevents using three separate, small transformers instead of one large transformer with three secondary windings? There certainly is enough space on the printed circuit board of LPS. In this case, the overall dimensions of LPS will differ not much more than SMPS. Even in the case of a powerful source with one level of an output voltage, it is possible to use some the flat transformers connected between themselves in parallel. So, the presence of the small transformer is not an absolute advantage of SMPS.

In connection with the SMPS operational ability over a very wide range of input voltages due to use of PWM in a control system of the switching element, this advantage is touted to us as the essential reason for selecting SMPS over LPS. Well, really, in practice is it important that SMPS can work at the input voltages changing within the limits of from 48 up to 312 V? In fact, this range comprises at once some rows of rated voltages, such as 48, 60, 110, 127, and 220 V. It is abundantly clear that in concrete equipment, the SMPS will work at any one rated voltage (changing within the limits of no more than ±20%), instead of on all of them simultaneously. And if it is necessary to use the equipment with both 110 and 220 V, for example, there are well-known solutions in the form of the small switch and tap winding of the transformer to handle this.

The coefficient of efficiency is an important parameter in the case of a powerful power supply, but not in the case of a power supply as small as 25–100 W, which we are considering. A high coefficient of efficiency and absence of a heat dissipation (characteristic of an SMPS) can be of some importance in miniature portable power supplies of completely closed systems—for example, as the power supply of laptops. In many other cases—for example, in controllers and electronic relays for industrial purposes—the coefficient of efficiency of power supply is not that important.

The possibility of functioning when fed from a DC network is the only major and absolute advantage of the SMPS. The LPS cannot operate in a DC mains power.

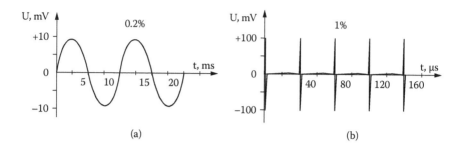

FIGURE 2.8
Typical levels of output voltage pulsations for (a) an LPS (0.2%) and (b) an SMPS (1%) with an output voltage of 12 V.

Here, in brief, we have given an analysis of the advantages of the SMPS in contrast to the LPS. We shall now turn our attention to the disadvantages of the SMPS.

One of the disadvantages of SMPSs is a high level of impulse noise on the power supply output (Figure 2.8). Unlike an LPS with its weak 50 Hz pulsation (Figure 2.8a), the pulsations of an output voltage in the SMPS (Figure 2.8b), as a rule, have much greater amplitude and cover a wide frequency range from several kilohertz up to several megahertz. This creates problems with high-frequency radiation that influence circuits within electronic equipment in which the SMPS is mounted.

These radiations also affect the wires that in turn affect electronic equipment that is external to the SMPS. Also, in SMPSs it is necessary to take steps to prevent the penetration of high-frequency radiation in the feeding power network (from which it is passed and can disturb the operation of other electronic devices) by using special filters (Figure 2.9).

The presence of a high-frequency component in the output voltage and in the internal SMPS circuits has led to increased requirements for the numerous electrolytic capacitors that are available in the internal SMPS circuits. Unfortunately, these requirements are seldom considered by engineers in the development process of the SMPS. As a rule, the types of these capacitors are selected only on the basis of their capacitance, operating voltage, and dimensions, without taking into account their high-frequency characteristics.

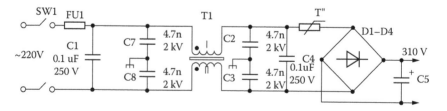

FIGURE 2.9
The circuit diagram of the typical input filter of the SMPS.

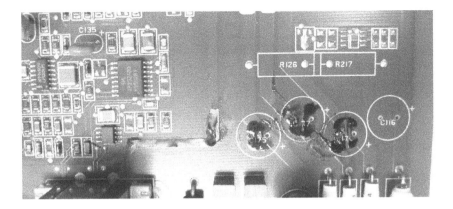

FIGURE 2.10
Destruction of copper streaks on the printed circuit board, which is taking place under capacitors because of electrolyte leakage.

However, not all types of capacitors have long life under the effect of a high-frequency voltage, but rather only special types having low impedance at high frequencies. As a result such nonsuitable electrolytic capacitors often heat up noticeably because of high dielectric losses at high frequency.

The rise in temperature of an electrolyte intensifies chemical reactions inside the capacitor that lead to a speed-up in the dissolution of the capacitor shell and even to an outflow of the electrolyte directly onto the printed circuit board. In very dense installations, this leads to shorting outlets of other elements or breakdown of circuits owing to the dissolution of copper paths of the printed circuit board (even despite presence of a strong covering of paths by a special mask; Figure 2.10).

Another well-known kind of SMPS fault caused at high temperatures of an electrolyte is the slow desiccation (over several years) of an electrolyte and significant (30%–70%) decrease of the capacitance that leads to a sharp decline in the characteristics of the power supply—and sometimes to full loss of its working capability [1].

For maintenance of effective work of the PFC, the power switching element (it is usually a MOSFET) should possess lower impedance in the conductive state. The value of this impedance largely depends on the maximum operating voltage of the transistor. For transistors with a maximum operating voltage of 500–600 V, this impedance achieves 0.05–0.3 Ω, whereas, for transistors operating at higher voltages (1000–1500 V), this impedance is one to two exponents higher (for example, 12 Ω for the transistor 2SK1794 for 900 V; 17 Ω for transistor IXTP05N100 for 1000 V; 7 Ω for transistor STP4N150 for 1500 V).

This is explained by selecting the low voltage (maximum operating voltage 500–600 V) transistors for the PFC. For example, in an actual SMPS design of such crucial devices as microprocessor protective relays and microprocessor-based emergency mode recorders, the following transistor types are widely used: IRF440, APT5025, etc., with the maximum operating voltage of 500 V. This is definitely not enough for functioning in an industrial electric network with a rated voltage of 220 V because of the presence of significant switching and atmospheric voltage spikes. As is known, for protection against such spikes, electronic equipment is usually supplied with the varistors. However, because of insufficient nonlinearity of the characteristic near an operating point, the varistors are selected so that between their normally applied operating voltage and a clamping voltage there will be essentially enough difference.

For example, for varistors of any type intended for operating at a rated voltage of 220 VAC (volt-ampere characteristic), the clamping voltage is 650–700 V. In the power supplies of microprocessor devices mentioned earlier, varistors of type 20K431 are used with the clamping voltage of 710 V. This means that at spikes with amplitudes below 700 V, the varistor will not provide protection for electronic components of the power supply, especially power transistors (500 V) that have been connected directly to a main network.

Both the transformer and the coil in PFC have high impedance at high frequencies that limits a current carried through them and through switching elements. However, a malfunction in the integral microcircuits providing control of the power switching element in PFC or basic power switching element of the SMPS (for example, as a result of an impulse spike) leads to a transition from high-frequency AC operating mode to DC mode (that is, with very low impedance) and a sharp current overloading of the power solid-state elements and to their instant failures. Considering the high density of the SMPS printed circuit board, this often leads to a fault of the nearby elements and burnout of whole sections of printed circuit strips. It should be absolutely clear that reliability of such a complex device as the SMPS, containing, as it does, an assemblage of complex microchips and power solid-state elements working at high voltage in a pulsing mode with high rate of current and voltage rise, will always be appreciable below the reliability of such a simple device as the LPS, containing only some electronic components that work in a linear mode.

The density of elements on the printed circuit board and specific power of the SMPS constantly increase; for example, an EMA212 type power supply (Figure 2.2, on the right, later in text), with dimensions as small as 12.7 × 7.62 × 3 cm, has an output power of 200 W. These are found in use in miniature electronic components having surface mount technology (SMT), with a very dense installation of high-power elements and constant increase of switching frequency. In the past, this frequency did not exceed 50–100 kHz. Today, many powerful SMPSs with output currents of up to 20 A operate at frequencies of 300–600 kHz; less powerful ones—for example, those controlled by an ADP1621 type integral circuit—operate at frequencies of 1 MHz and more. This promotes the further decrease of SMPS mass and dimensions. For the same purpose, the SMPS in an DPR is sometimes placed on a common board with other DPR devices—for example, with output relays (Figure 2.11).

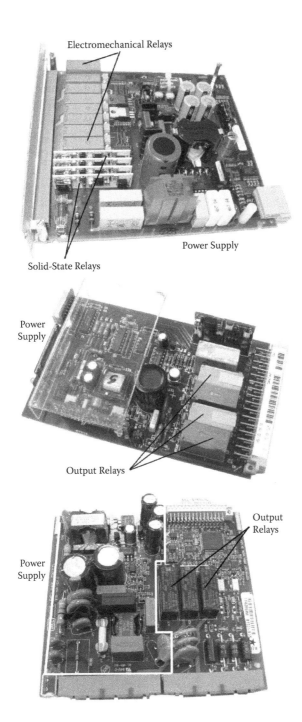

FIGURE 2.11

Power supplies installed on PCB common with other functional modules.

This tendency in SMPS evolution has been advertised as its great advantage. The downside of this is the practically full loss of maintainability of the SMPS.

However, are such power supplies necessary in the devices installed in the control cabinet? The question can be put even more broadly: Are the built-in power supplies in the electronic devices for industrial purposes intended for installation in control cabinets together with tens of other analogous devices necessary in general? Why not release in completely automatic systems (in the control cabinet) such devices as control units, electronic relays, and electronic transducers without power supplies, and only with a connector intended for connection to an external power supply?

This external power supply installed in the control cabinet (Figure 2.12) should be, in our opinion, linear to have a good power reserve and should be supplied by necessary elements for overvoltage and short-circuit protection. Moreover, in the control cabinets in reference to automatic systems with increased reliability, there should be two such LPSs and they may connect between themselves through a diode (a so-called "hot reserve"). It may seem strange but in the era of the SMPS, many companies (VXI, Lascar, Calex Electronics, Power One, HiTek Power, R3 Power, among many others) are continuing to manufacture LPSs. This testifies to its popularity in certain areas of techniques and its accessibility to practical applications. In our opinion, the approach specified previously would allow considerably increasing the reliability of the automatic systems such as remote control and relay protection

FIGURE 2.12
Cabinets with DPR installed.

(with a main supply of an alternating current) without increasing cost (owing to the smaller cost of electronic devices without the built-in power supplies).

An analogous approach can be used and, in the case of feeding the electronic equipment (for example, the same microprocessor protective relays) installed in the control cabinets from a network of a direct current, the only a difference is that two power supplies should be SMPSs instead of LPSs. Thus, these SMPSs should be subjected to serious redesign. First, the power factor correctors as absolutely senseless units for main supply from direct current should be dropped; this in itself will increase the reliability of SMPSs. Second, those SMPSs intended for installation in a control cabinet should be large enough (in such SMPSs it is senseless to pursue compactness) and convenient for repairing; they should not contain miniature SMT elements.

Third, the numerous electrolytic capacitors that are available in SMPSs should be concentrated on a simple separate printed circuit board intended for its simple replacement after each 10- to 12-year period of maintenance (that is, before the capacitors start to fail).

The main filter should be used as a complete device (hundreds of models of such filters are present in the market), instead of assembled from separate elements. Thus, it is possible to replace it simply and quickly if necessary.

In the some novel DPR of the last generation with integrated printed circuit board (PCB) modules, the power supply was made again as a separate module, as in early DPR constructions (Figure 2.13a, b). Such construction eases the idea's realization.

The solutions we offer, in our opinion, will allow lowering dependence of the stationary electronic industrial equipment (such as DPRs) on secondary power supplies and considerably increase its reliability.

In summary, some words about the newest trends that have appeared in the design of secondary sources of power supplies: the attempt to use microprocessors in LPSs [2] and also in SMPSs [3]. It may be that our view seems excessively conservative to the reader, but, in our opinion, microprocessors are necessary in power supplies as well as in toilet seats, where microprocessors are employed to measure temperature exactly of a matching part of a body and heat the seat up to a comfortable temperature.

It has become abundantly clear, as we have pointed out, that the presence of functionally unnecessary complex units in the equipment is the unequivocal way to decrease its reliability.

2.3 Problems with Electrolytic Capacitors

2.3.1 Introduction

Recent decades have been characterized by an explosion in computer technologies and the increasing use of switching power supplies; at the same

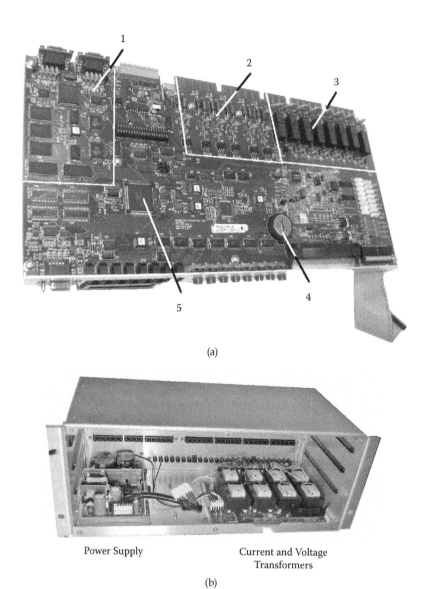

(a)

(b)

Power Supply   Current and Voltage
Transformers

FIGURE 2.13

(a) Universal PCB with all DPR modules (SEL-311L) except power supply and input trans-formers: 1—connection section; 2—digital input section; 3—output relay section; 4—bat-tery for memory backup; 5—main processor (highly integrated 32-bit microcontroller MC68332ACEH25 type). (b) Power supply and current/voltage transformers made as separate modules (SEL-311L, Schweitzer Electrical Laboratories).

time, failures of aluminum electrolytic capacitors have become so widespread that they have been dubbed a "capacitor plague" and have cost hundreds of millions of dollars. According to some published data, electrolytic capacitors have caused up to 70% of all damage to computers and computerized systems. Reasons for this situation have been mythologized and an incredible story has migrated from one magazine to another and from one Internet site to another. According to the story, in 1999 (or in 2001 according to other sources) an unspecified Chinese scientist working for a Japanese company engaged in the production of electrolytic capacitors and managed to steal the secret formula of the newest electrolyte. The problem was that the formula stolen was incomplete and now millions of electrolytic capacitors with a "terrible" water-based electrolyte have flooded the world. Within a few days or months, these capacitors absorb hydrogen from the air and explode, ruining the motherboard and any chip in which they are installed.

The authors of this anecdote ignored the fact that they are referring to capacitors from dozens of different manufacturers, including the Japanese, and that this problem has remained for the past 10 years. They "forget" about hundreds of patents on capacitor electrolytes registered in the patent collections of many countries, including the United States and Russia. Moreover, they do not care that such patents include detailed descriptions of the chemical composition and production technology of the electrolytes and that any chemical laboratory equipped with modern analytical equipment is able to determine the composition of the electrolyte taken from the capacitor. As one can see, stealing the formula of the electrolyte is senseless and this myth is a fake, apparently invented by journalists who were not very competent in this area. But, nonetheless, the problem really exists—and not just in computers. I found hundreds of damaged electrolytic capacitors of different types in power supply modules of dozens of failed microprocessor-based relay protection devices (MPDs) of different types and different manufacturers. So why has this problem been aggravated in the last decade? Let us try to understand.

2.3.2 Design Features of Aluminum Electrolytic Capacitors

First of all, let us look at the arrangement of the conventional aluminum electrolytic capacitor (Figure 2.14). As we see in the drawing, the design of an electrolytic capacitor is very similar to the design of the old paper capacitors. There are two layers of foil and one layer of paper between them, rolled and covered with protective aluminum housing. However, despite the similar appearance, there are fundamental differences in the design of electrolytic capacitors. The major one is that, unlike in a paper capacitor, in electrolytic capacitors, paper tapes are not used as the insulating material between the electrodes (plates) because they are saturated with conductive electrolyte and act as separators holding the liquid electrolyte in their pores. Between the plates there is a very thin insulating layer (its thickness is several fractions

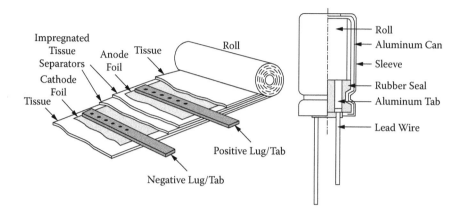

FIGURE 2.14
Design of aluminum electrolytic capacitor.

of microns) of aluminum oxide (Al_2O_3) covering the surface of the anode foil. Thanks to the small thickness of the dielectric (unattainable for capacitor paper), capacitors of this type have very large capacity (compared to paper capacitors), which is known to be inversely proportional to the distance between the plates. Increasing the area of plates (the surface area) adds to the capacitance. In the electrolytic capacitors, anode foil surface area is increased with electrochemical etching (before the formation of an oxide layer), after which the surface becomes somewhat rough (see Figure 2.15).

The greater the "roughness" is, the greater is the area. This method allows increasing the capacitance 20–100 times. The electrolyte, which is actually acting as a cathode, easily penetrates into the surface roughness of the foil. In such a capacitor, the cathode foil acts as a supportive electrode, connecting the electrolyte to the outer negative output. Sometimes the surface is etched to improve the contact with the electrolyte and reduce the contact resistance.

Aluminum oxide exists in several crystalline forms, the most common of which is α-Al_2O_3 or corundum, known in jewelry as ruby (containing red-colored impurities) and sapphire (with blue-colored impurities). This crystal is practically insoluble in water and in acids, and it is the *n*-type semiconductor

FIGURE 2.15
Surface of anode foil after electroetching.

forming the equivalent of a diode under physical contact with metals. (VAC of such contact is a typical characteristic of the diode.) This property of aluminum oxide determines the presence of diode D in the electrolytic capacitor equivalent circuit (see Figure 2.16). This diode is connected in the opposite direction and its breakdown voltage limits the operating voltage of the capacitor.

The same diode conditions need to comply with the polarity of the conventional electrolytic capacitors. Inductance of aluminum electrolytic capacitors (approximately 20–200 nH) is primarily determined by the inductance of the foil winding. It is usually not taken into account in the calculation of capacitor impedance, as the impedance of the capacitor is dominated by its equivalent series resistance (ESR) subjected to the resistance of the electrolyte and the outputs of the anode and cathode, including internal transient contact resistance.

However, this is true only for relatively low frequencies (below 100 to 1000 kHz). At high frequencies, the inductance markedly affects the impedance

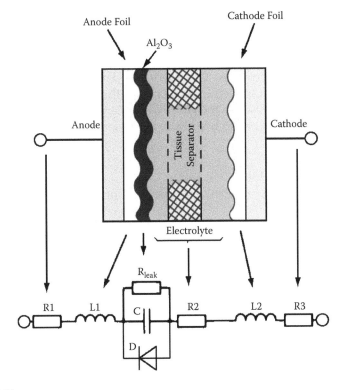

FIGURE 2.16
Equivalent circuit of an electrolytic capacitor: R1, R3—lead resistance on anode and cathode, including internal transient contact resistance; R2—electrolyte resistance; L1, L2—winding inductance of the anode and cathode foil; (R_{leak})—impedance of leakage through defects in the aluminum oxide layer; C—capacity of the aluminum oxide layer (capacitance); D—equivalent diode formed by a layer of aluminum oxide applied on aluminum foil.

of the capacitor; therefore, such factors as the equivalent series inductance (ESL) should also be considered. In fact, ESL limits the maximal operating frequency of the capacitor. The greater the ESL is, the lower is the limit frequency at which the capacitor has any capacitance. However, since the electrolytic capacitors are not designed to operate at high frequencies, the manufacturers of these capacitors rarely publish this value in the reference documents.

Aluminum oxide is a very hard and brittle material that can crack during rolling, cutting, or the operation of the capacitor, resulting in microcracks and micropores that can be penetrated with conductive electrolyte, increasing the leakage current. In addition, the coefficient of linear expansion of aluminum is several times greater than that of the oxide film, so changes of the temperature at the interface result in additional internal stresses, which may also lead to defects (cracks). The lower the leakage current is, the better is the capacitor. In good electrolytic capacitors, this current does not exceed tens to hundreds of microamps (depending on the size, temperature, and applied voltage). The chemical composition of the electrolyte must ensure recovery of the aluminum oxide layer to microdamage. And this is not the only requirement of the electrolyte.

Modern electrolytes for capacitors are complex multicomponent mixtures of acids and salts, in which the electric current flow is supported by ions and is accompanied by electrolysis. The electrolyte determines the efficiency of the capacitor under certain nominal voltages within a certain range of operating temperatures, as well as the nominal ripple current and the life of the capacitor. An operating electrolyte has to meet various and often conflicting requirements [4]:

- High intrinsic conductivity
- Small changes in conductivity over the entire range of operating temperatures
- Good formability: forming of anode (i.e., rapid recovery of the dielectric film of aluminum oxide on the edges and microcracks, formed during the cutting of foil and the winding of the capacitor element on the aluminum foil anode)
- Stable performance at the maximum operating temperature
- Lack of corrosion and chemical compatibility with aluminum, aluminum oxide, and capacitor paper of the separator
- Good wicking property of cushioning capacitor paper
- Stability of parameters during storage under normal conditions
- Low toxicity and flammability

The main components of the electrolyte are ion formation substances (ionogens), organic and inorganic acids, and their salts, but they are rarely used in their natural form. Typically, they are dissolved in a suitable solvent to produce electrolytic dissociation with the desired viscosity and formation of the electrolyte ions. Acids that can be used include monocarboxylic acids (nonane, oleic, stearic acids) and dicarboxylic acids (succinic, adipic, azelaic, sebacic, dodecane dicarboxylic, pentadecandioic acids), and phosphoric, boric, and benzoic acids (or ammonium benzoate). Boric acid enhances the forming ability of the electrolyte. For medium- and high-voltage capacitors, lactone and amide solvents can be used as a solvent. Electrolytes based on lactone solvents ensure high reliability and long service life of medium- to high-voltage capacitors, but the lower limit of operating temperature of such capacitors is limited, as a rule, to –55°C.

Electrolytes based on amide solvents ensure the lower limit of the capacitor operating temperature of –60°C or even lower. However, these electrolytes are not able to provide long service life for the capacitor, as they are very volatile and react with the aluminum oxide on the anode and destroy it, which leads to an increase in leakage current in the capacitor and a reduction of its service life. On the other hand, reduction of content of amide solvents and replacement of them with other solvents that are less volatile and less aggressive to aluminum oxide reduce low temperature characteristics of the electrolyte and, thus, of the capacitor along with the conductivity of the electrolyte. The electrolyte should not generate excessive gas (the operation of an electrolytic capacitor is accompanied by electrolysis resulting in the development of hydrogen at the cathode of the capacitor) at higher temperatures, including the top limit of the operating temperatures range.

Introduction of such additives as cathode depolarizers (e.g., aromatic nitro compounds) into the electrolyte enables a reduction of gas generation. Specific conductivity depends on the residual water content in the electrolyte including water generated from the chemical interaction of its components. Addition of deionized water can increase the electrical conductivity of the electrolyte. As a result, to meet all these requirements, the electrolyte becomes a sufficiently complex chemical compound consisting of many components, such as [5]:

- Ethylene glycol
- Alkanol
- Acetonitrile
- Sebacic acid
- Dodecanoic acid
- Ethyldiisopropylamine
- Boric acid
- Gipofosforistaya acid
- Ammonium hydroxide
- Deionized water

Finally, the parameters of the electrolyte depend on both its composition and mixing technology, while the capacitor electrical characteristics and service life largely depend on the parameters of its electrolyte.

During long-term operation of the capacitor, there are thousands of complex electrochemical reactions associated with the restoration of the oxide layer and with the corrosion attack to some internal elements, such as foil-electrode connection points. As a result of the inevitable corrosion processes, the ESR of the capacitor increases, leading to an additional increase in temperature and greater intensification of adverse chemical and physical processes inside the capacitor; this accelerates deterioration of its parameters. The process of natural increase in the ESR (i.e., natural aging of the capacitor) is rather slow (10–20 years and more).

In addition to aging, in some cases, premature failure of the capacitors takes place. The main reason for this is overheating. When the capacitor temperature reaches the boiling point of the electrolyte, the internal pressure increases and a certain amount of electrolyte goes out through the drain in the bottom plug or through the special valve (in large capacity capacitors), or through the special gap at the top of the aluminum cup (see Figure 2.17).

The ESR rises in proportion to the loss of electrolyte, resulting in further heating up. This positive feedback leads to a rapid capacitor failure.

Due to the loss of electrolyte, capacitance in electrolytic capacitors sharply decreases, sometimes accompanied with a complete break of the internal circuit. What is going on in electronic equipment during electrolyte drain?

First, a significant decrease in the capacitance affects the normal operation of many circuits: The filtering of the variable component is impaired, voltage on sensitive circuit elements is reduced, etc. Evidence of DPR usage suggests cases of mass failure of relay types SPAC, SPAU, and SPAJ (manufactured by ABB) due to a significant reduction in the capacity of a single capacitor of 100 μF in the power supply unit (see Figure 2.18).

FIGURE 2.17
Electrolyte drain paths in aluminum capacitors: 1—special notch attenuating the bottom of an aluminum cup; 2—plastic or rubber glass covering the plug and fixing the outputs; 3—valve in the high capacity capacitors.

100 uF, 50 V

SPGU240A1

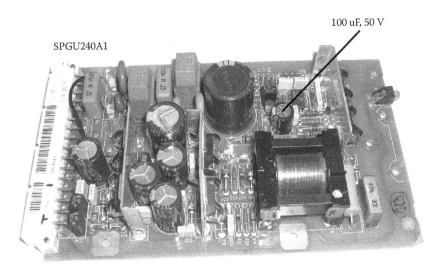

100 uF, 50 V

SPTU240S1

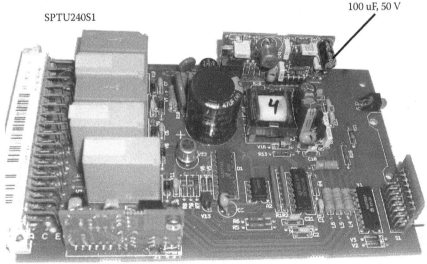

FIGURE 2.18
Power supply units of types SPGU240A1 and SPTU240S1 of microprocessor-based protective
relays type SPAC, SPAU, and SPAJ (ABB).

Second, contact with conductive electrolyte causes short circuiting and
failure of the microelectronic components' outputs. If electrolyte contacts
the power supply components that are under line voltage, the power circuit is short-circuited accompanied by intense arcing and explosive physical
destruction of these elements and emission of a large amount of electrically
conductive soot onto adjacent components (see Figure 2.19).

Furthermore, acids contained in the electrolyte rapidly destroy the varnish
coating of printed circuit board (the mask) and dissolve copper tracks on the

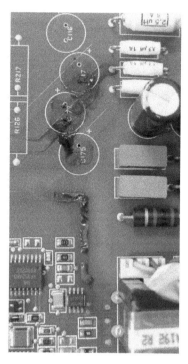

FIGURE 2.19
Destruction of PCBs and elements due to contact with electrolyte leaked from capacitors.

PCB (see Figure 2.19). Sometimes, as temperature and pressure grow, the electrolytes of certain composition demonstrate faster loss due to evaporation of volatile fractions through the plug rather than due to leakage. Occasionally, usage of poor quality electrolyte causes internal chemical reactions in the capacitor with emission of large amounts of hydrogen that leaks through the plug seal. In such cases, the amount of electrolyte in the capacitor is also reduced (it partly gasifies) along with its capacitance, which can go down 10-fold within 5–10 years.

What causes premature failure of aluminum capacitors? Undoubtedly, the poor quality capacitors made in violation of the processes from the poor quality materials will not last long in the equipment. However, let us dismiss the incompetent theory of a "stolen" bad recipe mentioned previously. Its worthlessness has been shown. It should be noted also that power supply units contain quite a few capacitors of the same type included in various circuits but that failure occurs in only one of them (see Figure 2.18) or in a group included in a particular circuit (see Figure 2.20).

Marked groups of capacitors can cause massive power failure due to leakage of electrolyte. This directly implies some other theory and another cause behind mass failures of capacitors. Analysis of circuits containing electrolytic capacitors that experience frequent failures shows that we are dealing

FIGURE 2.20
Power supply type 316NN63 of microprocessor-based relays series RE * _316 (ABB).

here with circuits operating under high-frequency voltages (used in switching power supplies). State-of-the-art high-power switching power supplies operate at frequencies of tens of kilohertz and low-power ones in the range of hundreds of kilohertz [6].

Since the dielectric losses (dissipation factor: *tan* δ; *tan* δ = 2p*f*C*R*, where *R* ≈ *ESR*) are directly proportional to the frequency *f*, it is clear that additional losses occurring at these frequencies cause further heating of the electrolyte and, hence, an increase of pressure inside the capacitor, with all the consequences that come with it. However, as we can see from the preceding formula, the losses in the capacitor added to its heating are directly proportional both to the frequency and to the ESR. And this means that opting for extremely low ESR capacitors in switching power supplies may essentially reduce electrolyte heating and extend service life of capacitors as rated in the manufacturers' technical manuals for operation under maximum allowable temperature. Thus, for K50-75 type capacitors, mean time to failure (MTTF) at +85°C shall be no more than 1,000 hours, while reducing the temperature to +55°C results in a longer operational time of up to 10,000 hours [7].

It should be noted in this context that the method suggested by some authors for damage protection of electrolytic capacitors, such as bypassing by small-capacity ceramic capacitors, is a common misconception. At frequencies of tens to hundreds of kilohertz, impedance of small-capacity ceramic capacitors by far exceeds even the worst electrolytic capacitor ESR. But in order to protect the electrolytic capacitor from the effects of these frequencies effectively, the protective capacitor ESR should be at least comparable to that of the capacitor to be protected. A simple calculation shows that to meet this condition, the capacity of protective capacitor at a frequency of 100 kHz should be about 5 µF, and this is characteristic of a big film capacitor with high PCB space requirements, rendering this solution unacceptable.

Subject to the standard IEC 60384-4-1 [8], technical documentation for the oxide electrolytic capacitors should reference their impedance at a certain frequency. Impedance in international practice is usually referenced at 100 kHz, typical frequency for switching power supplies. At this frequency, impedance and the ESR are virtually the same. Technical manuals of Western manufacturers of low ESR oxide electrolytic capacitors may name capacitor series as follows: low impedance, very low impedance, ultralow impedance, extremely low impedance. Analysis of impedance values for these capacitor series shows that actual figures correlate with the superlative degrees in their series names only on rare occasions. Therefore, such names should be regarded as an advertising gimmick only and should not be relied upon.Whenever one chooses electrolytic capacitors to be used in switching power supply, one should check the impedance of the capacitor at a frequency of 100 kHz against the manufacturer's technical documentation. Unfortunately, quite often technical manuals of Russian manufacturers do not reference any impedance values for common, general purpose capacitors at all. For some capacitor types (such as K50-75, A DPK.673541.011 TУ) available in 33 group sizes, the impedance value is referenced only for four of them. And even in respect to military purpose capacitors (index acceptance: "5")—classified in the low impedance group (such as capacitor types K50-83, АЖЯР.673541.012 TУ)—technical manuals do reference the value of ESR and impedance, giving no frequency and temperature at which the value is guaranteed. Therefore, valid evaluation of these specifications cannot be made. And only for a very limited number of capacitor types produced in Russia do technical manuals clearly and accurately reflect the impedance value, making it possible to compare them with the world's top capacitors (see Table 2.1).

TABLE 2.1

Impedance of Oxide Electrolytic Capacitors Made by Leading World Manufacturers for Frequency of 100 kHz and Temperature of 20°C

FM (Panasonic) for capacitors:	ZL (Rubicon)	HD (Nichicon)	KZE (Nippon Chemicon)	Voronez Capacitor Factory K50–38	Voronez Capacitor Factory K50–53	Impedance for Frequency 100 kHz at Temperature 20°C
0.038	0.053	0.053	0.053	0.3	0.4	6.3 V, 1000 µF
0.026	0.041	0.038	0.038	0.35	0.3	25 V, 470 µF
0.061	0.12	0.074	0.074	0.6	1.0	50 V, 100 µF

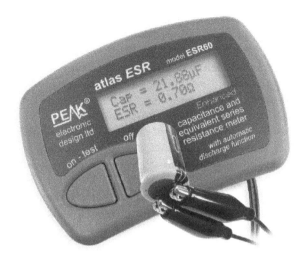

FIGURE 2.21
Best-selling ESR60 type device used to measure capacitance and equivalent series resistance (ESR).

The data in Table 2.1 clearly show that Russian manufacturers still have a lot of room to improve the parameters of their capacitors.

Capacitor ESR can be assessed based on manufacturers' technical manuals and measured directly by simple devices operating at the standard frequency of 100 kHz. Several models of such simple and relatively low cost devices (USD 150–200) are available in the market—for example, ESR60 manufactured by Peak Electronic Design (Figure 2.21), which can be purchased through the global distributors of electronic components, such as RS or Farnell.

In most commercially available devices of this type, the health of capacitors may be accessed directly in the circuit without desoldering them.

It should be noted that the reliability of switching capacitors and capacitor-input filters also depends on the maximum allowable ripple current. Ripple current flowing through the electrolyte further heats it, and a condenser operating at the upper limit of the allowable temperature range has a very short life, usually up to 1,000 to 2,000 hours. When selecting an electrolytic capacitor, it is therefore important to consider this characteristic, which usually is contained in the manufacturers' manuals.

The evolution and ever wider application of microprocessor devices have uncovered another problem related to electrolytic capacitors. Today's powerful processors constitute the so-called dynamic load and operate in a pulsed high-frequency mode of consumption of rather high currents in power circuits. Traditional computer processors consume current of 5–10 A. In state-of-the-art powerful processors with billions of transistors (Intel four-core processor known as Tukwila contains over two billion transistors, and their number in a new NVIDIA Fermi graphics processor already exceeds three billion), input current reaches some tens of amperes. This means that in the processor power circuits, the capacitors will be exposed to significant high-frequency charging and discharging currents, which is no better than the operating conditions in switched power supplies. Therefore, massive failures of electrolytic capacitors are not limited to the power supplies only. They occur in motherboards and processor supply circuits as well.

The good news is that, unlike primary power supplies, state-of-the-art high-performance microprocessors operate at very low voltages. Thus, while the first-ever microprocessors operated at a supply voltage of 5 V, the latest generation of microprocessors has much lower voltage requirements. Thus, the Intel® Xeon® processor can operate at 1.5–1.33 V while consuming current of up to 65 A, which makes it possible to use surface-mounted low voltage capacitors of other types (other than aluminum oxide capacitors designed for voltages of up to 600 V) on the motherboard.

The most popular alternative to aluminum oxide capacitors has been presented by tantalum capacitors. These capacitors are believed to outperform aluminum ones far and away because they are the capacitors used in special purpose military and aerospace equipment. But is this actually the case? What are tantalum capacitors?

2.3.3 Design Features of Tantalum Electrolytic Capacitors

There are at least two large classes of tantalum capacitors: one with liquid electrolyte and one with solid electrolyte. The main difference in design between tantalum capacitors and aluminum capacitors is their respective anode and cathode design. Unlike aluminum oxide capacitors with anode made in the form of tape coiled into a roll, the anode in both classes of tantalum capacitors is designed in the form of a highly porous three-dimensional cylindrical tablet made of tantalum powder pellets sintered in vacuum at 1300°C–2000°C with the wire lead pressed in from the inside (Figure 2.22).

These capacitors utilized the ability of tantalum to form (by electrochemical oxidation) the oxide film on its surface: pentoxide tantalum Ta_2O_5, a highly stable, high-temperature compound resistant to acidic electrolytes and conducting current in one direction only, from the electrolyte to the metal. The electrical resistivity of pentoxide tantalum film in the nonconducting direction is very high (7.5 10^{12} Ωcm). This anode design determined the name of the USSR's first series of tantalum capacitors: ЭTO-1 and ЭTO-2 (ETP-1 and ETP-2—electrolytic tantalum porous—in English translation). Their commercial production was launched in the late 1950s to early 1960s (see Figure 2.23).

These capacitors proved to be so good that, despite their half-century of age, they are still produced by Oxide Novosibirsk plant branded K52-2 (ОЖО.464.049 ТУ) and with acceptance index "5" and "9" (that is, made to military and space requirements).

FIGURE 2.22
The structure of the sintered tantalum pellets.

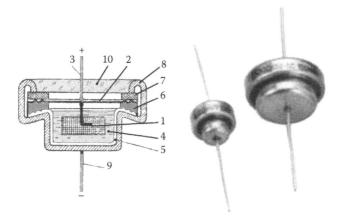

FIGURE 2.23
The design of ETP series tantalum capacitor and its modern counterpart of K52-2 series: 1—tantalum anode; 2—tantalum cap; 3—anode lead; 4—electrolyte; 5—inner silver shell; 6—cushion; 7—insulating gasket; 8—outer steel casing; 9—cathode lead; 10—epoxy sealing.

These capacitors usually use 35% to 38% aqueous solution of sulfuric acid (H_2SO_4) as a working electrolyte. It is this concentration of sulfuric acid that ensures its maximum conductivity and the lowest freezing point (about −60°C). Sulfuric acid–based electrolyte used in capacitors ensures resistivity of about 1 Ωcm at 20°C. Less aggressive electrolytes were suggested earlier, but they have higher resistivity (i.e., ESR: H_3PO_4 solution—4.8 Ωcm, LiCl solution—12 Ωcm, etc.), so they are not widely used.

The presence of an aggressive electrolyte such as sulfuric acid necessitates the use of double casing: an inner thin-walled silver shell (neutral to acid)

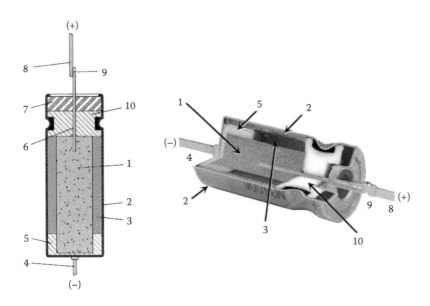

FIGURE 2.24
The design of state-of-the-art tantalum capacitor with liquid electrolyte: 1—tablet made of the sintered tantalum pellets; 2—silver (silver plated) shell–cathode; 3—electrolyte (acid); 4—cathode lead; 5—inside Teflon insulator; 6—anode lead made of tantalum wire; 7—insulation plug (occasionally, glass insulator); 8—anode lead (tin-plated nickel); 9—welding point of anode leads; 10—PTFE wall tube.

and an outer stainless steel casing providing sufficient mechanical strength. Great attention has to be paid also to the design sealing to prevent possible leakage.

Modern tantalum capacitors with liquid electrolyte are not essentially unlike the samples that were launched 50 years ago, but they have a cylindrical form, more familiar to modern capacitors (see Figure 2.24).

The second class of tantalum capacitors features solid electrolyte. As follows from the very name of this class of capacitors, their main difference from the previous ones is the absence of liquid electrolyte. These capacitors are also called oxide-semiconductor (solid-electrolytic) capacitors because they use manganese dioxide (MnO_2) as a solid electrolyte known to have semiconducting properties. A layer of manganese dioxide atop the tablet made of pressed tantalum pellets with a premanufactured pentoxide tantalum layer is formed by keeping it in a manganese nitrate solution followed by drying at a temperature of about 250°C. This creates a manganese dioxide layer that is used as the capacitor cathode.

Mechanical and electrical contact of the outer lead with the manganese dioxide layer is achieved as follows: The manganese dioxide layer is coated with graphite; the graphite, in turn, is covered with a layer of silver to which a wire cathode lead is soldered. The cathode lead of a casing intended for

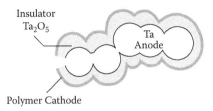

Insulator
Ta_2O_5

Ta
Anode

Polymer Cathode

FIGURE 2.25
The structure of solid-state tantalum capacitor with polymer electrolyte.

surface mounting is made of an electrically conductive epoxy compound (with powdered silver filling).

Recent years have brought to life various types of tantalum capacitors with solid electrolyte differing in the composition and technology of conductive layer application to the tablet made of pressed tantalum pellets.

Most notably, solid electrolytes based on conductive polymer have proliferated (see Figure 2.25). There are several types of conductive polymers that have found use in tantalum capacitors:

- Tetracyano-quinodimethane (TCNQ)
- Polyaniline (PANI)
- Polypyrole (PPY)
- Polyethelyne-dioxythiophene (PEDOT)

The latter type of polymer has found the most practical use for the manufacture of capacitors (and much more).

Tantalum capacitors with solid electrolyte are free from the serious flaws of aluminum oxide capacitors such as electrolyte drying and leakage. But let us take a closer look at some of the characteristics of tantalum electrolytic capacitors. Having said that tantalum capacitors certainly outperform aluminum oxide ones, it should come as some surprise to know that the ESR, this critical characteristic, is by far worse in tantalum capacitors with liquid electrolyte compared to traditional aluminum capacitors (see Figure 2.26); that, unlike aluminum capacitors with their maximum operating voltage of up to 600 V, maximum voltage of tantalum capacitors is limited to 125 V (and for most types even to 50 V); that tantalum capacitors fail to withstand the slightest overvoltage and even short voltage surges not exceeding their maximum allowable values results in breakdowns with shorting the circuit in which they operate. Breakdown and current flow result in strong heating of the capacitor and the release of oxygen from manganese dioxide; taken together, they cause a violent reaction of oxidation and inflammation of the capacitor, which can set equipment on fire. To prevent breakdown of tantalum capacitors and to extend their life, they are used at voltages two to four times lower than maximum allowable ratings. Given the fact that no

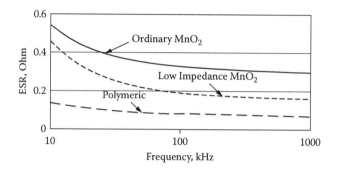

FIGURE 2.26

Relationship between equivalent series resistance (ESR) and frequency for different types of capacitors with solid electrolyte.

TABLE 2.2

Typical ESR Values at 100 kHz for Tantalum Capacitors at Temperature of 20°C

Type and Manufacturer	Nominal Voltage, V	Capacitance, µF	ESR for Frequency 100 kHz, Ω
Solid tantalum 293D series, Vishay Intertechnology, Inc	6.3	1,000	0.4
	20	100	0.5
	50	15	0.8
M3900622H0190, Cornell Dubilier	100	22	0.4

tantalum capacitors are available for voltages exceeding 125 V (bulk production is intended for voltages of up to 50 V), application of such capacitors is rather limited.

Specified ESR (impedance) values are compared in Tables 2.1 and 2.2 for aluminum oxide and tantalum electrolytic capacitors.

Also, tantalum capacitors are much more expensive as compared to aluminum ones. And even special types of tantalum capacitors claimed to be low ESR capacitors still fall far behind the best types of aluminum oxide capacitors (see Table 2.3).

A positive touch to this grim picture is introduced by the fact that tantalum capacitors with polymer cathode are less flammable than capacitors containing manganese dioxide, and they have lower ESR values (see Figure 2.26).

But why, after all, are tantalum capacitors so good? Why are these capacitor types used in military equipment?

All types of tantalum capacitors have lower leakage currents, longer life, and, more importantly, much wider operating temperature range than aluminum oxide capacitors. For example, K52-18 tantalum capacitors have minimum life of 150,000 hours at 0.6 of rated voltage and temperature of +55°C. Their operating temperature ranges from −60°C to +125°C and beyond (for example, +155°C for K52 series), which fully satisfies the requirements of

TABLE 2.3

Typical ESR Values for Low-Impedance Tantalum Capacitors at 100 kHz and Temperature of 20°C

Type and Manufacturer	Nominal Voltage, V	Capacitance, µF	ESR for Frequency 100 kHz, Ω
Solid tantalum, TRS series, Vishay Intertechnology, Inc	6.3	1000	0.1
	20	150	0.1
	50	15	0.3
CWR29 series, AVX	6	330	0.18
	20	47	0.11
	50	4.7	0.5

Russian Military Standard GOST PB 20.39.304-98 to environmental conditions for military equipment, but has no particular importance for industrial applications (e.g., in digital protective relays with a much narrower range of operating temperatures).

Recently, high-capacity (100 µF and beyond) multilayer ceramic capacitors have been developed that are free from many of the shortcomings typical of electrolytic capacitors. However, the capacity of these capacitors is still highly dependent on temperature, they are significantly more expensive than electrolytic capacitors, and they are not yet widely accepted.

2.4 Conclusions and Recommendations

1. The main characteristic of electrolytic capacitors that should be considered for the development of new switching power supplies or repair of failed units is the equivalent series resistance or impedance at the frequency of 100 kHz, which must have minimum values.

2. Electrolytic capacitor protection from high-frequency components through bypassing small capacity ceramic capacitors is inefficient at frequencies used in switching power supplies.

3. State-of-the-art microprocessor operation is accompanied by the consumption of significant currents in high-frequency pulse mode, so the capacitors placed in the power circuits of microprocessors are exposed to high-frequency charging and discharging currents. For this operation mode, one should also choose the capacitors with minimum ESR value.

4. The main types of damage in aluminum oxide electrolytic capacitors for switching power supplies are the drying up or leaking of

electrolyte accompanied by dramatic decrease in capacitance, disruption of the supply unit operation, and damage caused to PCB components by leaked electrolyte.

5. The main types of damage in tantalum capacitors for central processor units are breakdowns accompanied by shorting the circuit in which they operate.

6. Comparative analysis of the characteristics of aluminum oxide capacitors versus tantalum capacitors has revealed that contrary to a common misconception about the absolute qualitative supremacy of tantalum capacitors, they fall far behind aluminum capacitors in terms of such important characteristics as ESR. In addition, tantalum capacitors operate at a much narrower range of voltages, which is clearly insufficient for switching power supplies, and fail to withstand even minimal overvoltage.

7. Commercial switching power supplies are better served by aluminum oxide electrolytic capacitors. The circuits wherein the capacitors may be exposed to high frequencies should use special types of capacitors with low ESR. In this case, one should be guided by the data referenced in technical manuals or measurements made by special tools rather than by advertising names of such capacitors. For such applications, most suitable capacitors are series FM, KZE, HD, and ZL.

8. Tantalum capacitors with solid electrolyte intended for surface mounting have smaller dimensions than aluminum capacitors and are more widely accepted and more convenient for use in CPU units. But they, too, should be chosen based on the minimum value of ESR if intended for microprocessor power circuits and with 200% to 300% rated voltage.

9. In order to prevent unexpected and fatal damage to switching power supply units operating in critical electronic equipment including digital protective relays manufactured 7 to 10 years ago, it is advisable to get them examined, to identify numbers of damaged capacitors, and proactively to replace these capacitors in all power supply units before they fail, keeping in mind the recommended guidelines suggested in this chapter. While doing so, with old capacitors soldered out, their mounting locations on the printed circuit board and leaked electrolyte traces should be washed with sodium bicarbonate solution and then with distilled water and dried thoroughly.

References

1. Gurevich, V. 2008. Reliability of microprocessor-based relay protection devices—Myths and reality. *Engineer IT* Part I: 5:55–59; Part II 7:56–60.
2. Sadikov, Y. 2008. Power supply adapter with output voltage adjustment in the interval 1.5 to 15 V and output current up to 1 A. *Electronics-Info* 12:42–43
3. EFE-300/EFE-400. 300/400 W, digital power solution. Datasheet TDK-Lambda, 2009.
4. The working electrolyte for a capacitor, method of its preparation and aluminum capacitor with this electrolyte. Russian Patent No. 2358348, H01G9/-35, 2006, "Elekond" Plant OJSC.
5. Conductive electrolyte system with viscosity reducing co-solvents. US Patent No. 6744619, H01G 9/42, 2004, Pacesetter, Inc.
6. Gurevich, V. I. 2009. The secondary power supplies: Anatomy and application. *Electrotechnical Market* 1 (25): 50–54.
7. Russian Technical Specification # АДПК.673541.011 ТУ. К50-75 Oxide-electrolytic aluminum capacitors.
8. IEC 60384-4-1. Fixed capacitors for use in electronic equipment—Part 4-1: Blank detail specification—Fixed aluminum electrolytic capacitors with non-solid electrolyte—Assessment level EZ.

3

Battery Chargers

3.1 Purpose and Modes of Operation of Battery Chargers

If we consider only the name—battery chargers (BCs)—we can make the erroneous assumption that BCs are only intended to charge the accumulator batteries and maintain them charged. But this is not true. BCs represent a main power source of auxiliary circuits under the normal mode of operation of a direct current auxiliary power system (DCAPS) at a substation or power plant. While the battery is activated, the BC cannot supply enough energy to DCAPS consumers. This happens in the absence of power in the AC auxiliary power, when the BC is inoperable or when the BC is not powerful enough to cover rare peak loads of the DCAPS.

The BC specifications have to ensure their efficiency under different modes of operation with an automatic switch between them. According to reference 1, the following are the main modes of operation of a BC:

1. *Floating charging*—The battery is charged with a low current in order to compensate for self-discharge and maintain it in a fully charged condition.

2. *Equalizing charging*—The battery is charged in order to balance voltage on battery cells.

3. *Boost charging*—Unfortunately, reference 1 provides neither definition nor explanation of this mode. But it is known that it is used for a rapid return of the battery to normal operational condition after full discharge as well as for partial cleaning of electrodes from sulfation.

In the condition of floating charging, the BC has to ensure consistent output voltage on the battery of 2.15–2.23 V per cell; in equalizing charging, 2.3–2.4 V per cell; and in boost charging, up to 2.7 V per cell. Precise voltage depends on the type of the battery and is usually indicated by the manufacturer in the manual accompanying the accumulators. The choice between different modes of battery charging and switching between them in a BC can be either manual or automatic (based on a timer), while switching between

BC modes of operation in the course of charging a discharged battery is done only automatically in all types of modern BCs.

According to reference 1, BCs must ensure the charging of a battery in a three-stage automatic mode:

- Stage I: limiting initial charging current at $0.3\,C_{10}$
- Stage II: limiting charging voltage
- Stage II: a mode of voltage stabilization

where C_{10} is the capacitance of a battery equivalent to a 10-hour discharge cycle.

As Figure 3.1 shows, the BC works in a mode of current limiting during the first stage. This mode of operation is achieved by the automatic reduction of the BC's output voltage to the level at which the current rate does not exceed a previously set value.

Limiting the maximum current rate is necessary not only for a battery, but also for a BC, since internal resistance of a discharged battery is too low, which can result in overloading and breakage or immediate disconnection of its output automatic circuit breaker in the absence of the current limiting function of a BC. Different types of BCs used in the power sector have difference rates of outgoing current: from 10–15 A to 300–500 A. This is the level at which charging current of a battery is limited, though it can be adjusted to some extent by BC adjustment mechanisms. This is the mode when accelerated and the most intensive charge of a battery is taking place, which is accompanied by intensive chemical reactions and gas emissions—particularly hydrogen, which can be explosive if mixed with oxygen.

At some point of time in the course of battery charging, the voltage starts increasing. Increasing antielectromotance of a battery results in reduction of the current rate, which is taken from a BC. At some point, this rate becomes lower than the maximum current rate set in a BC. A BC exits the current limiting mode and enters the mode of voltage limiting. This is when

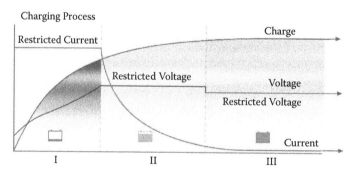

FIGURE 3.1
Illustration of charging process of discharged battery by an automatic three-stage BC.

additional charging of a battery takes place, which is accompanied by a slow increase of its charge and gradual reduction of a current rate consumed by a battery. Finally, when a certain minimal level of current is achieved, the BC switches to low voltage and enters the mode of voltage stabilization in which the current rate consumed by a battery does not exceed a few tens to hundreds of milliamperes, which are used for balancing the self-discharging of the battery.

In order for this current to flow from BC to battery, output voltage of the BC should be a bit higher than the voltage of an idling battery. When the DCAPS load takes current lower than that preset by the current limiting of the BC, this current is consumed from the BC exclusively, rather than from a battery. The latter is automatically activated when its voltage becomes higher than the output voltage of the BC (as a result of BC current limiting starting). Further distribution of load current between the battery and the BC will be proportional to the levels of their voltages.

It often happens that after stage I, the BC automatically switches to stage III. Such BCs are called two-stage devices. There is no big difference whether there is a stage II, but the process of battery recovery after its discharge runs more quickly if this stage is present. There are also four-stage BCs on the market, where the charging current is preset with gradual increments rather than in steps at the initial stage of charging.

3.2 Arrangement and Principle of Operation of a Classical BC on Thyristors

As one may understand from the previous section, a BC is a device that serves as a rectifier, current stabilizer, voltage regulator, and voltage stabilizer. This list can be extended by a function of galvanic separation of circuits of operating supply of AC/DC, as well as a function of changing the ratio of output to input voltage rate. BCs usually have a reduced (e.g., 125, 60, or 48 V) output voltage compared with input voltage or, alternatively, increased output voltage (e.g., 400–800 V, which is used for battery charging in high-power uninterruptable power supply [UPS]).

Let us review the flow diagram of the power part of a BC based on a simple but reliable device with an output current rate of 30 A and nominal voltage of 230 V. Though this device was manufactured in the 1970s–1980s by Winterfeld, it is still efficiently working in many substations. These devices are much simpler than modern tools based on microprocessors; that is why it is easier to study the principle of the BC's operation on these devices.

The majority of BCs' circuits are constructed on power thyristors that are working at industrial frequencies. The mode of operation of a halfway,

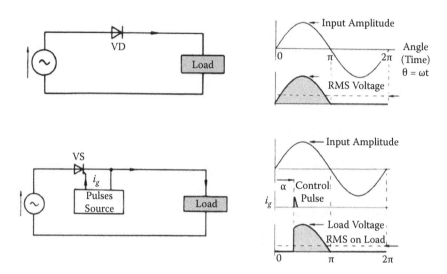

FIGURE 3.2
Principle of operation of half-wave circuits: above—on a diode (VD); below—on a thyristor (VS).

one-phase, thyristor-based circuit rectifier is shown in Figure 3.2. The difference between a rectifier on a thyristor and a rectifier on a diode is that there is a full half-wave of AC on the port of a diode rectifier, but only a part of this half-wave on the port of a thyristor rectifier, since the thyristor (unlike a diode) does not conduct current until it is brought into conductor state by means of a control impulse, which is delivered to a controlling electrode (gate) of the thyristor. When this control impulse is delivered to the gate with a shift (delay) compared with applied voltage phase, we obtain a cutoff half-wave of voltage. By changing the area of a half-wave of output voltage by means of changing the phase of the control impulse, we can adjust the r.m.s. rate of output rectified voltage. Using a bridge layout with four thyristors (full bridge) or two thyristors and two diodes (half-bridge), it is possible to obtain a Graetz circuit with rectified adjustable voltage (see Figure 3.3).

Using a three-phase bridge rectifying circuit (Figure 3.4) enables solving several problems simultaneously:

- Reducing asymmetric load of phases
- Improving the form of output voltage by reducing pulsation of output voltage (AC component of voltage or current at the output port of rectifier)
- Greatly reducing the level of harmonic components, which are delivered by the rectifier into an AC supply circuit by using a transformer at the input port with winding connected in delta-star scheme

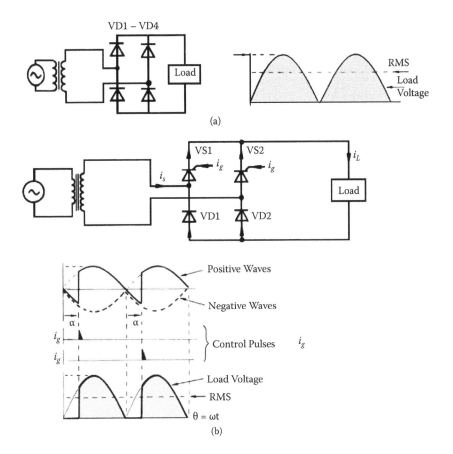

FIGURE 3.3
Principle of operation of half-wave circuits on (a) diodes and (b) thyristors.

When using thyristors instead of VD1–VD6 diodes, we obtain a three-phase rectifier voltage regulator as shown in Figure 3.4. These are the three-phase rectifier voltage regulators that are often used in powerful industrial BCs for currents from 30 to 100 A (Figure 3.5). For devices with input currents of 300–500 A, six-phase rectifier voltage regulators on thyristors or two separate power transformers are used: one transformer with a secondary "star" layout winding and the other with "delta" layout winding. Each of them is working the ports that are connected parallel utilizing its own three-phase thyristor rectifier regulator. This allows reducing the level of pulsation of rectified voltage, when other approaches (e.g., use of bulk capacitor) are inefficient at high load current.

Figure 3.5 shows a full-flow diagram of Winterfeld's BC, which was manufactured in the 1970s and 1980s. We decided to use this relatively outdated device as an example, as it is relatively simple and very reliable. (They are still efficiently working on many substations.)

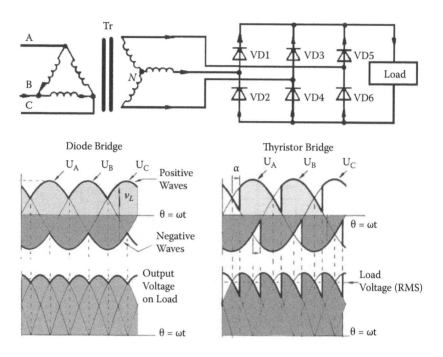

FIGURE 3.4
Principle of operation of a three-phase bridge rectifier on diodes and three-phase rectifier-voltage regulator on thyristors.

In this device, three-phase alternating voltage comes to the power transformer (Tr), through the input automatic circuit breaker (CB) and contactor K. There are three current transformers (CTs) and a filter (Ftr) on the side of low voltage of this transformer. Alternating voltage from the secondary winding of the transformer is delivered to the full three-phase rectifying bridge on thyristors VS1–VS6.

Rectified voltage is filtered by chokes L1 and L2 and a high-capacity capacitor C. There is an activated load resistance PR behind the filter, which is used to create small loads for thyristors when the BC is working in the idle mode or in a mode of very low current. In the absence of this load (i.e., in the idle mode), the work of the thyristor regulator is unstable. Further, there is a shunt circuit in the DC line, which is used to measure output current of the BC and an output automatic circuit breaker (DC CB). The thyristors are protected from transition overvoltages by means of RC circuits (capacitor connected in series with resistor). Gates of the thyristors are connected through impulse transformers (IT), with a control unit CU2 (Figure 3.6), which, in turn, is connected with a control unit CU1 (Figure 3.7). These universal control units were developed and manufactured by the AEG Company for different types of BCs. It may be surprising, but the same control units are installed in one of the most modern models of BC for 300 A current, manufactured by Benning.

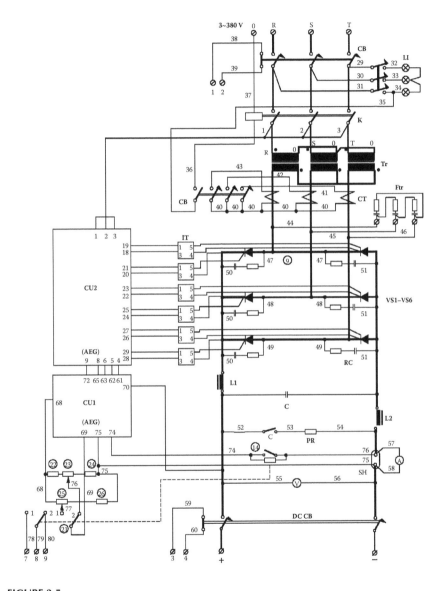

FIGURE 3.5
Circuit diagram of Winterfeld's BC.

Outputs to Pulse Transformers

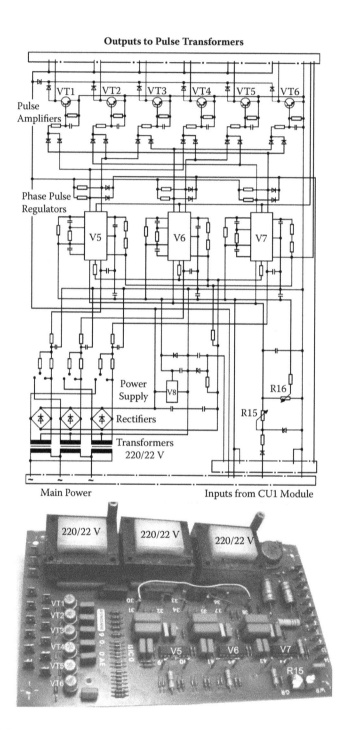

FIGURE 3.6
Circuit diagram and arrangement of an impulse control block CU2.

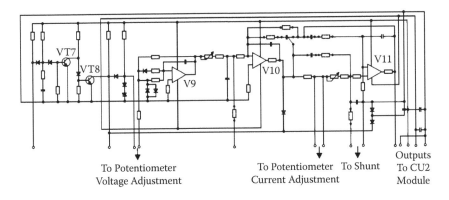

To Potentiometer Voltage Adjustment

To Potentiometer Current Adjustment

To Shunt

Outputs To CU2 Module

FIGURE 3.7
Circuit diagram and arrangement of an analog control block CU1.

The ports of control unit CU1 are connected to potentiometers, which adjust the level of output voltage under modes of floating and equalizing charging, as well as shunt circuit terminals, included in a circuit of the BC's output current. This unit has been built based on two transistors, VT7 and VT8, and three operational amplifiers, V9–V11 series 741. It produces some resulting voltage, which is proportional to the position of potentiometers and the rate of output current, measured by the shunt circuit. This resulting signal comes to the port of the impulse control unit CU2. By means of V5–V7 chips type UAA145 in this unit, impulses for thyristor control are generated, which are amplified by transistors VT1–VT6 and come to the ports of the ITs and on to the gates of thyristors. The impulses that are generated by V5–V7 chips are synchronous by phase with three-phase voltage of supply line from transformers 220/22 V. Rectifying bridges, connected to secondary winding of transformers, together with voltage stabilizers V8 are intended to provide power to the elements of the CU2 unit. The analogue signal, coming from the CU1 unit, results in a shifting of the phase of signals generated by V5–V7 chips relative to the phase voltage of the supply line. As mentioned previously, this shift of signals delivered to gates of the thyristors results in changing the RMS rate of output voltage of the BC (and its output current under the mode of current limiting).

Very high-power chargers with microprocessor-based controllers are also available now in the market. Such chargers consist of a large circuit breaker (1), bus bars, two power transformers: one Star-connection (2) and another Delta-connection (3) each with its own rectifier (4) connected in parallel for best smoothing of output-rectified voltage. Such chargers for output current of 1200 A (for example, Figure 3.8) are about 12 m in length.

3.3 BCs with a Function of Discharging the Battery

The correct use of batteries at substations and power plants includes not only correct charging, but also occasional discharging of the battery. Actually, in order to measure the real capacity of the battery, it needs to be fully discharged. Moreover, a process of natural aging of the battery is slowed down if it is occasionally fully discharged with a subsequent charging. This is especially important for batteries that are continuously working under the mode of floating charging. This is why many BCs are equipped with a kit of powerful resistors intended for battery discharge by means of a controlled current. More advanced models ensure discharge of the battery into the supply line (i.e., they return energy accumulated by the battery—a process called recuperation). The main difference of these models from the simple BC described earlier is that they have additional terminals in the output line of the BC (Figure 3.9), which enables changing the polarity of the BC's battery when it is switched into discharge mode. These models also have an additional controller (Figure 3.10), which ensures synchronization with the line and series control of each thyristor according to a specific algorithm.

It is known that thyristors cannot switch into the nonconducting state on their own (to shut down) when direct current is constantly flowing through them, the rate of which is higher than some very low rate of maintaining current. However, the opening of a specific thyristor in the circuit of a BC under the mode of battery discharge results in application of voltage that is inverse to voltage coming to the thyristor, through which discharging current is flowing at the moment. As a result, this conducting thyristor shuts down and switches to a nonconducting state. Thus, by alternating the opening of

FIGURE 3.8
High-power doubled charger for output current 1200 A and 230 V.

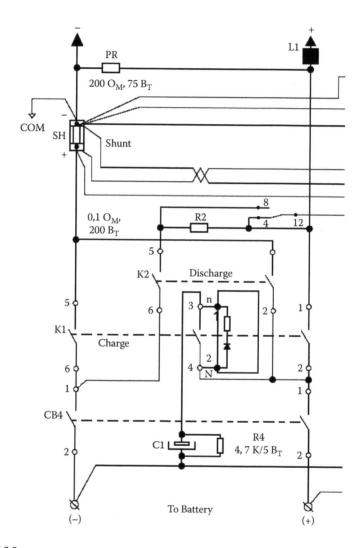

FIGURE 3.9

Fragment of diagram of power part of BC with additional contactors K1 and K2, which are used to switch from charging mode into discharging mode.

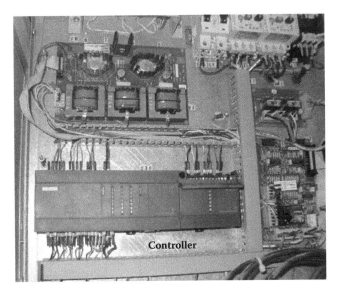

FIGURE 3.10
Controller type SIMATIC S7-200 (Siemens), used in BC of ELCO company to control the process of the battery discharging into the supply line.

shutdown thyristors in a specific sequence, it is possible to shut down other open thyristors.

As a result, a thyristor circuit creates rectangular impulses of current with opposite polarity (meander), which are converted into sinusoids after passing through choke inductance and the windings of the power transformer (which works in this mode as if in a reverse direction) and are delivered into the line in this form.

Taking into consideration that the process of battery discharging should run under a constant rate of current, it becomes obvious that it is not easy to establish a necessary algorithm of BC operation and that this requires a separate controller. The more advanced models of BCs with microprocessor control do not require separate controllers.

3.4 BCs with Two Rates of Output Voltage

For mobile substations of 160 kV, ELCO manufactures an original BC on thyristors that has two rates of output voltage at the output ports: one rate to supply a system of DC buses (60 V, 20 A) and the other to charge the battery (80 V, 24 A) (Figure 3.11). This necessity is explained by lack of room for a big stationary set of batteries.

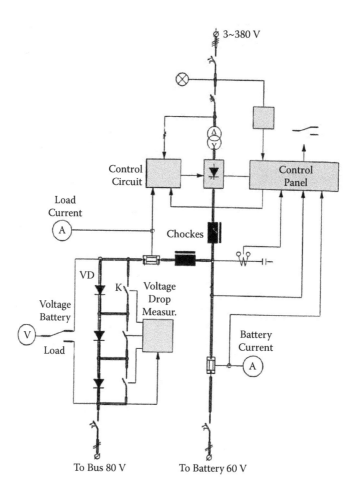

FIGURE 3.11

Flow diagram of BC with two rates of output voltage.

In this situation it is possible to increase the energy capacity of the battery by means of increasing its voltage through a connection of a large number of batteries in series. In this case the battery consists of six 12 V batteries (like those in a car) connected in series. This battery series, charged to about 80 V, is capable of supplying energy to auxiliary circuits with a nominal voltage of 60 V much longer than a battery with the same capacity but with fewer batteries and with 60 V.

Two rates of output voltage in this device are ensured by original nonlinear resistance obtained from a large number of VD diodes connected in series in the forward direction and K contactors shunting them (Figure 3.12). A thyristor rectifier regulator in this device maintains only the output voltage (80 V) on the battery at the preset level. Stabilization of voltage on 60 V buses

FIGURE 3.12
Design of BC with two rates of output voltage.

under variable load current is performed with a help of nonlinear resistance on the VD diodes.

Fine stabilization of output voltage on buses is achieved due to nonlinearity of volt-ampere specification of diodes. Rough stabilization is achieved by shunting of separate diodes with the K contactors based on the signal of the layout of measuring voltage drop on diodes.

3.5 Principle of Action of BC with a High-Frequency Link

Not only permanent cabinet types but also mobile BCs are used in the power industry. These mobile BCs are mainly used as backup BCs (e.g., to maintain the battery's charge in case of breakage of permanent BC or during maintenance of a battery disconnected from DC power supply). Stationary BCs are not suitable for this because of their size and weight. Different companies used to try to manufacture smaller BCs for 15–20 A current and 230 V (see Figure 3.13), but they did not gain popularity due to their heavy weight, reaching 60–70 kg. The weight was so high because of the size and weight of the power transformer and smoothing choke, which operated at a frequency of 50 Hz.

There is a way to reduce the size and weight of chokes and transformers significantly by switching them to higher frequencies (tens to hundreds of kilohertz) at which the transformer of the same power can be much smaller both in size and weight. This approach is used in modern mobile BCs (Figure 3.14).

FIGURE 3.13
Mobile BC for 16 A current and nominal voltage of 230 V.

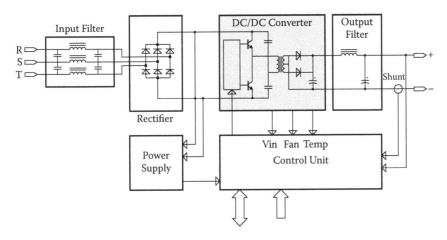

FIGURE 3.14
Structural diagram of a BC with an additional high-frequency link (converter).

Power convertors, which convert voltage of direct current of one rate into voltage of direct current of another rate, are usually installed on a powerful field or insulated gate bipolar transistors (IGBTs) that are working in switching mode (i.e., which can be opened and closed with a frequency of tens to hundreds kilohertz. Converters installed in a half-bridge layout (Figure 3.15a) or a full-bridge layout (Figure 3.15b) are used most often.

In the half-bridge layout (Figure 3.15a) capacitors C1 and C2 are charged to a half of power source voltage, each ($U_{C1,C2} = 0.5$ E). When transistor 1 is opened, capacitor C1 is discharged through the primary winding of transformer T. This impulse of discharging current flows through primary winding of a transformer and creates a voltage in the secondary winding. When transistor 1 is closed and transistor 2 is opened, the discharge current of capacitor C2 flows through the same winding of the transformer, but in the reverse direction.

Thus, impulses of current with opposite polarity (alternating current) flow through the primary winding of transformer T. Strictly speaking, transistors do not create only one impulse; they create a bunch of impulses of variable width. The control of the rate of output voltage is performed by changing the width of impulses that control transistors (i.e., by changing the time during which transistors are working in conducting state). This principle of transistor management is called pulse-width modulation (PWM; Figure 3.16). Due to the transformer's inductivity, groups of impulses of opposite polarity current (created by the transistors) start resembling ordinary sinusoids by form. Alternating voltage created in the secondary winding of the transformer is rectified, filtered, and delivered to the battery.

In the full-bridge layout (Figure 3.15b) two additional transistors are used instead of two capacitors. In this layout, a series of sequential openings and closings of transistor pairs (1 and 3, and then 2 and 4) results in

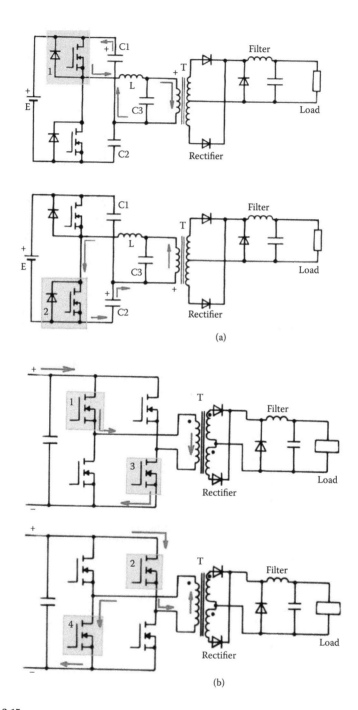

FIGURE 3.15
Explanation of principle of operation of converter in (a) half-bridge layout or (b) full-bridge layout.

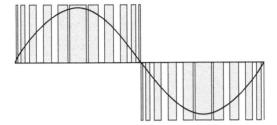

FIGURE 3.16
Principle of pulse-width modulation (PWM).

the creation of bipolar groups of current impulses in the primary winding of transformer T. The rest in this layout is similar to what was discussed for the half-bridge layout.

3.6 Ferroresonant-Type BCs

The main element of ferroresonant-type BCs is a ferroresonant stabilizer of voltage, which belongs to the ferromagnetic resonant type of stabilizer. Any stabilizer of this type includes linear ballast resistance Z_{LIN} connected in series and nonlinear element Z_{NL} with a load Z_H activated parallel to it (Figure 3.17).

In this device, when input voltage U_{IN} increases, resistance of a nonlinear element Z_{NL} decreases, which increases the rate of current flowing through it. As a result, current flowing through ballast element Z_{LIN} also increases and, consequently, the voltage drop on this element also increases. The specification of a nonlinear element is chosen in such a way that an additional drop of voltage on the ballast resistance compensates for voltage increase at the entrance of a system. In this case, the output voltage will remain unchanged.

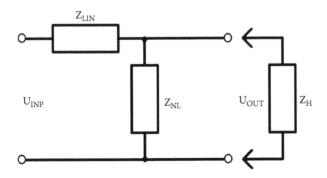

FIGURE 3.17
Arrangement of ferromagnetic resonant stabilizer of voltage.

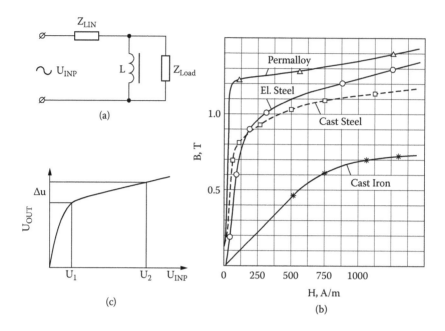

FIGURE 3.18
Principle of operation of ferromagnetic voltage stabilizer: (a) circuit; (b) magnetization curves of typical ferromagnetic materials; (c) dependence of output voltage on input voltage in ferromagnetic stabilizer.

Many semiconductor devices, as well as induced coils with saturated ferromagnetic cores (Figure 3.18), possess nonlinear characteristics. When working on the AC, the choke's core, L, will occasionally be saturated, and then the change of average output voltage Δu during the half-wave will be a bit lower than the voltage change at the input terminals of the circuit $u_2 - u_1$ (Figure 3.18c).

This happens because when input voltage increases and the choke's core is saturated, its inductive reactance is decreased and the rate of current flowing through it increases. This results in an increase of voltage drop at ballast resistance Z_{LIN}. At the same time, the voltage on the load terminal does not change, which means that there is an effect of voltage stabilization similar to any other parametric stabilizer.

Figure 3.18(c) shows that the nonlinear working part of such a stabilizer is far from ideal (horizontal) and this device has a rather limited working range of input voltage at which a necessary level of stabilization is achieved. Also, this layout will not have a high level of efficiency due to high rates of reactive current.

In order to increase the nonlinear characteristic of the circuit and to reduce the losses, another capacitor, C, is connected in parallel to a nonlinear choke, L, which creates a resonant circuit together with inductance L adjusted to the frequency of a power source (Figure 3.19a). In this circuit (which is called a

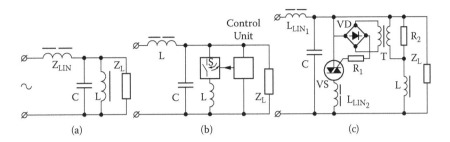

FIGURE 3.19
A principle of construction of (a) ferroresonant and (b, c) quasi-ferroresonant voltage stabilizers.

circuit with ferroresonance of currents), voltage stabilization starts at lower current rates than in the absence of ferroresonance.

The magnetizing choke's current L when closing the LC circuit enables reducing the reactive current flowing through Z_{LIN} and improves power characteristics of the circuit. As mentioned in Bogdanov [2], it is possible to create an oscillatory process in the LC circuit and efficiently use the energy accumulated in the condenser even without ferroresonance by artificial commutation of L inductance, which can be linear and can contain no core (Figure 3.19b). It is convenient to use a TRIAC VS (Figure 3.19c), the angle of opening of which changes with the change of input voltage, creating an analogy of a nonlinearity of saturated ferromagnetic core [2].

By introducing an additional change of angle of the triac opening, one can adjust actual rate of output voltage of such a stabilizer. Indeed, this device is similar in its characteristics to those of a ferroresonant stabilizer, although it is not a ferroresonant stabilizer as Bogdanov [2] suggests and thus it could possibly be called "quasi-ferroresonant."

In BCs used in practice, the resonant condenser as well as a resonant choke with a TRIAC are usually connected into the lines of the additional winding of a transformer (not parallel; Figure 3.20). The role of serial linear resistance is performed by the winding of this transformer. With this connection, the full load current does not flow through a TRIAC, which is the case in thyristor-type BCs, and its rate is much lower than the rate of load. Actually, the TRIAC in this layout is connected into a control circuit and not into a power line.

In other words, we can say that a BC of this type does not contain semi-conductor elements in the power line. This feature of a ferroresonant BC (quasi-ferroresonant would be a better name in this case, but it is not used in technical literature) is mentioned in promotional brochures of BC manufac-turers as a very valuable feature that can significantly increase reliability of the BC. Indeed, the absence of semiconductor elements in the power part and a simple control scheme tend to increase reliability of such BCs. Practical use of such devices confirms this.

The drawbacks of ferroresonant BCs may include the large amount of metal and weight of these devices, which consist of an electromagnetic circuit

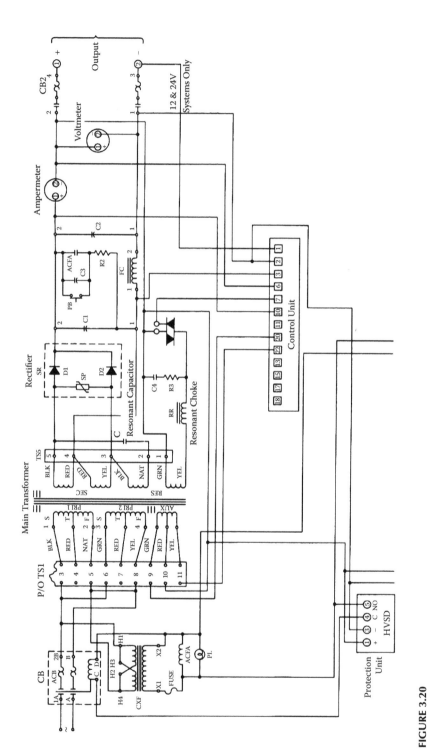

FIGURE 3.20

Circuit diagram of commercial BC of ferroresonant type series ARE. Manufactured by C&D Company.

FIGURE 3.21
Arrangement of one-phase ferroresonant BC series Battery-Mate of Ametek, Inc. Company (USA) with a 10-year warranty on ferroresonant transformers

made on a powerful multiwinding transformer (Figure 3.21). For example, a BC series "R" manufactured by Ametek (very similar to the BC depicted in Figure 3.21) weighs 290 kg in a single-phase, 90 A arrangement and 365 kg in a three-phase, 90 A arrangement. It is often difficult to match the small size of such BCs (Figure 3.21) with their huge weight. In one situation, four workers dismantling a relatively small cabinet (weighing several hundred kilograms!) hanging on a heavy-duty bracket dropped a BC after taking it off the bracket because they did not expect that a cabinet of this small size could weigh so much.

At the end of a review of different types of BCs, it is worth mentioning that the most widespread type of BC is a device with a three-phase, thyristor-based adjustable rectifier. Lightweight mobile BCs are manufactured with a high-frequency link. The most reliable and long-lasting BCs are those of a ferroresonant type, but they are also the heaviest.

3.7 Autoreclosing of a BC

Most industrial electrical installations, including power electronic devices (battery chargers, invertors, rectifying devices, UPSs, etc.) are protected from overload and internal short circuits by means of automatic circuit breakers connected on the side of power supply line. The features of these CBs are selected considering the overload capacity of internal elements so that the maximum level of protection efficiency is ensured. However, there have been situations where the practical application of power electronic devices shows that CBs are sometimes activated when there is no internal breakage or overload. If people are always on duty near such an electrical installation, there will not be significant problems: It is enough to switch the CB on again and that is it. But what happens if the substation is unattended or the problem happens at night?

Special research of such faulty activation of CBs has not provided an absolutely precise or unambiguous reply applicable to all situations. However, some interesting features of power electronic devices have been revealed in terms of chargers and controllable rectifier invertors (with energy recuperation). It appears that the reason of faulty activation of CB of such devices is the magnetizing current (inrush current) of the internal power transformer when it is switched on when it is idle. (This switching is preset by the internal logic of the device's operation.) For example, one of the three-phase chargers with an output voltage of 220 V and output current of 30 A showed a current inrush in the circuit of a switch reaching as high as 700–900 A. The other device of a similar type did not show current inrush exceeding several hundred amperes. The difference between these two charges lies in the distance of these devices from the external auxiliary transformer—in other words, in the different resistance of the supply line, which limits maximum current.

Of course, the duration of impulses of current with such amplitudes is not significant; however, activation of such a CB can be unavoidable under certain conditions (e.g., unfavorable combination of the CB's parameters, latching AC current phase, and supply line resistance). Unfavorable conditions often arise under emergency conditions in supply lines that are accompanied by a series of short-time voltage drops and recovery.

It is possible to avoid the effect of magnetizing current inrush onto a CB by using the simplest current limiter. Some types of chargers have a main contactor in series with a CB, which is activated during switching-on with a certain delay (up to several seconds) and that is necessary to prepare the internal controller for operating. In such devices, the current limiter is represented by the simplest structure (Figure 3.22), which includes three resistors connected parallel (by phases) to the main contactor's terminals. When the CB is on and contactor is off, a low current passes through the internal power transformer, limited by these resistors (0.1–0.2 A); this is enough for preliminary magnetizing of the transformer during the contactor's delay time. After the contactor is on, these resistors are shunted by its terminals and have no further effect on the operation of the device. Research with full-size devices has shown that these resistors may have resistance of 500–1000 Ω and power of 25–50 W.

Unfortunately, a transformer's magnetizing current inrush is not the only reason for the activation of CB, and this is why a solution on the basis of starting current limiting is not always effective. Research has shown that one of the reasons for such activation may be a transition process related to emergency situations in the high-voltage networks. At the same time, one can observe short-term shifts of phases and asymmetry of phase power in a 400 V circuit. A short-term failure is observed in the operation of electronic

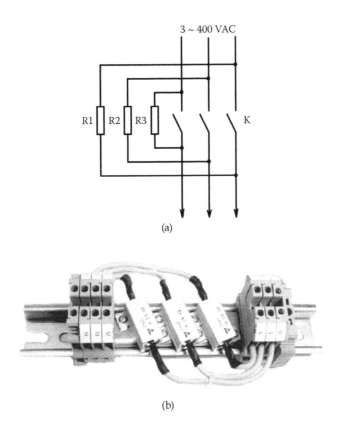

(a)

(b)

FIGURE 3.22
A set of three resistors intended for limiting of inrush current (magnetizing current of power transformer) and the circuit of its connection. K—main contactor.

control systems that control power semiconductor elements (thyristors, IGBTs), which constantly watch for an angle of phase shift. This situation results in simultaneous opening of semiconductor elements that should not be opened simultaneously. As a result, a short circuit appears, leading to activation of a CB. It is obvious that a limiter of inrush current will not help in this situation.

The most comprehensive solution here, which does not depend on a reason of faulty switching off the CB, is a use of the autoreclosing of electric installation. The basis of this device should be represented by an automatic CB with parameters similar to those of the main CB, since overload and short-circuit protection of an electric installation should be on under any mode. In the simplest case it is possible to use the same CB furnished with a special drive, which returns it to initial (switched on) condition. Such a drive (type FW7) is manufactured by the Moeller company (Figure 3.23).

This drive has a protruding push rod, which is connected to a handle of the switch and lifts it (switches on) with a certain delay after automaticly

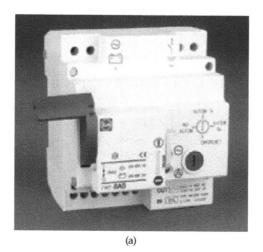

(a)

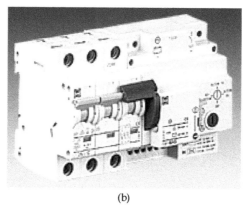

(b)

FIGURE 3.23

Type FW7 drive (Moeller) used for automatically switching on the CB. Right: the drive with a switch.

switching off. It is obvious that a device of this type cannot be universal since its sizes should strictly match the size of the CB. Also, button-type switches cannot be switched by means of such a drive at all. Moreover, the price of such a device (about USD 500) does not stimulate its widespread use.

In connection with the previous discussion, we have developed a simple device (Figure 3.24) suitable for electric installations of any type. The device consists of an additional automatic switch CB1″ of the same type as the standard switch CB1 installed in the electric device, an additional contactor K2, auxiliary relay K1, and a timer with a delayed switching-on. When the CB1 is on, a coil of the relay K1 receives energy through normally closed contacts of the timer. Relay K1 is constantly on and its contact is open.

When CB1 switches off spontaneously, relay K1 loses its power supply; its contacts close and supply power to the timer. After the several minutes

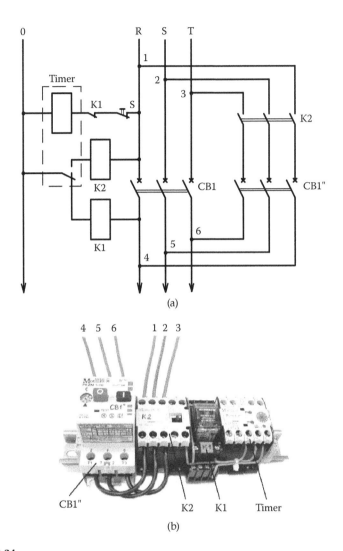

FIGURE 3.24
Autoreclosing device: (a) circuit diagram; (b) appearance.

necessary to finish all transition processes in the circuit, the timer is actu-ated and its contacts additionally disconnect the supply line of relay K1 and transfer power to a coil of contactor K2. A consumer is automatically con-nected through the contactor K2 and initially preswitched on CB", bypassing a disconnected switch CB1.

In order to return a device to its initial condition it is necessary to stop the power supply to the device from the main control cabinet for a short period of time and return the main CB1 to its initial condition (switched on). If short-term disconnection of the device is undesirable, one can introduce a return button S. Elements of any type with coils for 220 V AC can be used in the device. The power of the contactor should match the power of the electric installation.

It should be noted that particular care should be taken when servicing electric installations with autoreclosing. These installations should be marked with labels (Figure 3.25) requiring disconnecting of external power supply during maintenance or repair.

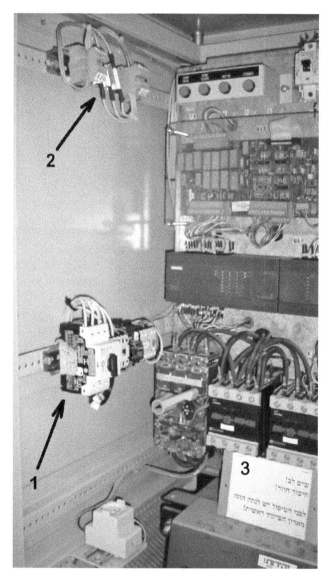

FIGURE 3.25
(1) Autoreclosing device; (2) inrush current limiter; and (3) a warning sign about availability of reclosing function in BC (ELCO Company).

More than 25 kits of the described devices (current limiter equipped with autoreclosing device) have been used in powerful chargers manufactured by ELCO.

3.8 A Problem of High Capacitance on BC Output Ports

To ensure a required quality of output voltage, high-capacity electrolyte capacitors are installed on BC output ports as filters. In modern BCs, the capacity of these capacitors can reach as much as several hundred thousand microfarads. Fault currents through such capacitors can reach several amperes if a large number of harmonic waves is present; this results in severe heating of the capacitors and an increase of fault current flow, which may explode a capacitor, create an internal short circuit in the power part of a BC, and damage to many power elements of a BC. Sometimes it can lead to a breakdown of the internal capacitor's insulation due to overvoltage in case of switching of inductive load in a DC line. To protect internal power elements of a BC, high-capacity capacitors are protected by special fuses of high disconnecting capability and a special signal terminal, which blocks operation of a BC in case the fuse is blown (Figure 3.26). However, when output CB of a BC is turned on, the battery's voltage is applied to the output capacitors one at a time. In this case, an impulse of charging current with large amplitude is flowing through the noncharged capacitors, which leads to blowing of the previously mentioned fuse and the blocking of the whole BC.

Also, a charging current impulse of large amplitude flows in the line of capacitors when the lead-in main CB of the BC is on from the side of the AC; this is why the changes in the mode of BC energize are not helpful.

An additional problem is imposed by the ability of the capacitors to keep the charge for a long time after the BC is disconnected. This increases the risk of electric shock and impairs maintenance of the BC.

It is not permissible to use fuses for current rates higher than the rating prescribed by a manufacturer. During some emergency modes of operation of the electronic control unit of BCs, there may be a situation when one of the thyristors stops opening, resulting in a sharp increase of harmonic levels consumed by a capacitor. As a result, the current rate flowing through the capacitors can reach 15–20 A. It will start heating and can explode the short-circuiting output circuits of the BC. This creates considerable damage for a BC. A 6 A fuse in the circuit of the capacitors used in this case will serve to protect a BC from such situations.

In order to prevent the fuse blowing and ensure safety during maintenance, we introduced two buttons and a resistor rating of 100 Ω, 50 W (Figures 3.27 and 3.28) in the BCs, which tend to have such problems.

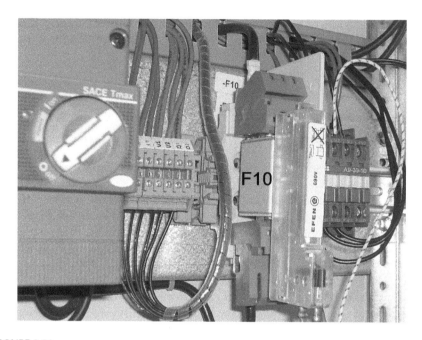

FIGURE 3.26
Special fuse F10, installed in the circuit of a high-capacity capacitor with a signal terminal, included in the BC's control circuit (Benning Company).

Before switching a BC on with a disconnected main CB, pressing the charging button will perform initial charging of a capacitor from the battery for 5–10 seconds and then switch on the main CB. This manual charging is required only in case of manually switching on a BC and is not required when the BC returns to operation after the AC supply has recovered after its loss, since condensers remain connected to the battery and maintain their charge. During maintenance, when the BC is turned on and off several times, condensers are discharged by a discharging button when the BC is turned off. Its own internal resistor R20 is used for this, though a charging resistor R can also be used.

3.9 Arrangement of Alarm about Failure of Transformers in Control Unit of BC

In order to connect control circuits with the supply line, miniature encapsulated power transformers are used in control units of BCs of different types. The overload capacity of such transformers is extremely low. Unfortunately,

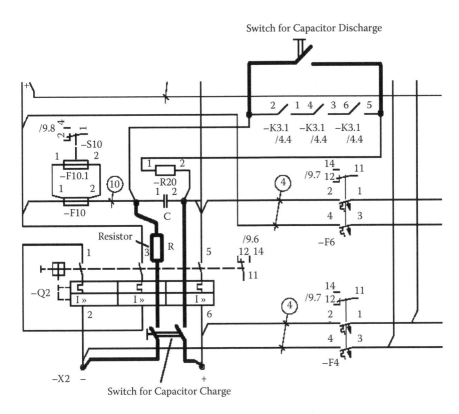

Switch for Capacitor Discharge

FIGURE 3.27
A fragment of a circuit of power, part of Benning's BC with additional charging/discharging buttons of condenser C and resistor R.

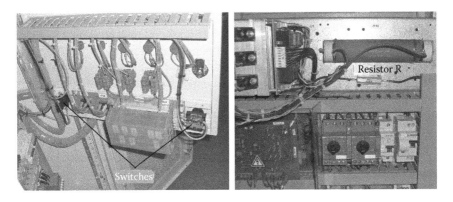

FIGURE 3.28
Additional buttons and a resistor, installed in Benning's BC.

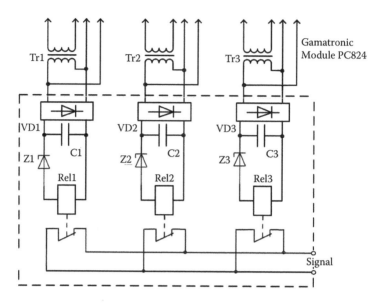

FIGURE 3.29
Circuit diagram of alarm unit about damage of miniature transformers (Tr1–Tr3) in the BC's module PC 824 manufactured by Gamatronic.

the lacquer insulation of coil microwires is so thin that they break very often. The damaging of the transformers results in the malfunction of the BC and, oddly enough, one gets no alarm from the BC about the malfunction. It is obvious that this situation is dangerous and intolerable. We cannot state that this is standard for all types of BCs, but several of them (e.g., BCs of Gamatronic Company) really do have this problem.

In order to solve this problem, a simple unit was developed that controls the functionality of the transformers and delivers a signal about their malfunction (Figure 3.29). This unit consists of three rectifying bridges, VD1–VD3; bulk capacitors, C1–C3; Zener diodes, Z1–Z3; and miniature highly sensitive relays Rel1–Rel3 type T81HD312-24 with coil resistance of 2.8 kΩ for 24 V. This device is assembled on a small plate and installed in the BC's cabinet (Figure 3.30). It controls the availability of output voltage on secondary winding of transformers Tr1–Tr3 and delivers a signal if the voltage is lost on any of the transformers. Due to the high sensitivity of the relays Rel1–Rel3, they do not create an additional load for transformers.

3.10 Problems of Electromagnetic Relays in BCs

Many electromagnetic relays are used in the control circuit of a BC. They transfer commands and control signals and holds between different

FIGURE 3.30
Arrangement of an alarm about damage of transformers Tr1–Tr3 in the control system of Gamatronic BC.

functional units of a BC. Contacts of several of these relays are used for switching of 220 V of AC/DC and several of them are used for switching of 5–12 V and current rate of several milliamperes. As a rule, manufacturers of BCs use only one type of relay in their products, which is chosen based on maximum switching current and voltage (Figure 3.31). At the same time, it is thought that if a relay can switch 8 A current at 250 V, it will definitely be able to switch 5 mA current at 5 V.

This is the reason one type of relay is used in BC. This allows a manufacturer to reduce the range of component parts assortment and reduce costs by purchasing of large batches of similar parts. But to what extent is it possible to use general types of commercial relays with powerful contacts for switching low-powered signals with low levels of voltage and current (so-called "dry" circuits)?

It appears to be impossible. That is why conscientious manufacturers of electromagnetic relays indicate in their specifications both the maximum and minimum ratings of switching current and voltage rates. Both maximum and minimum switching current and voltage rates depend on many structural factors and conditions, as do which relays will be used; they are reviewed in detail in Gurevich [3]. In particular, they depend on forms and sizes of contacts, contact pressure, and, what is even more important, the material from which contacts are manufactured. In order to increase the switching ability of small contacts of modern miniature electromagnetic relays for general purpose, such alloys of silver with nickel (AgNi), cadmium (AgCdO), and tin ($AgSnO_2$) are often used as contact material. These alloys

(a)

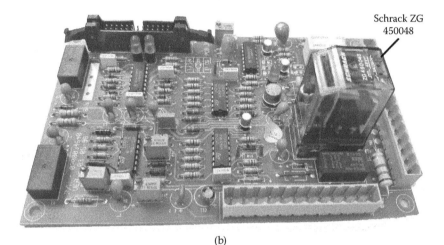

Schrack ZG
450048

(b)

FIGURE 3.31
(a) Units of electromagnetic relays, series Finder 40.52 (8 A, 250 V), used in Gamatronic BCs; (b) one of control units of ELCO's BC with Schrack ZG 450048 relay (5 A, 220 V), used to switch low-current signals.

are very resistant to electric arcing and provide high switching character-istics as well as a long-term resource even for small contacts of miniature relays, which are working in commercial automatic devices.

However, silver and silver-based alloys tend to oxidize and create noncon-ducting sulfide films on the surface. These films are quickly destroyed under high temperatures of electric arc and, thus, they have almost no influence on the operation of a relay. However, in the absence of an electric arc and with the absence of multiple switching, nothing interferes with film formation; this leads to the significant increase of resistance of contacts so that they dis-tort low-rate signals between electronic circuits inside the BC, resulting in its malfunction. This can happen after 5–10 years, depending on air humidity and sulfur and other pollutants' concentration in the air. Moreover, it is very difficult to find such a source of the BC's malfunctioning.

As mentioned earlier, some serious manufacturers indicate minimum rates of switching current and voltage in their specifications. For example, Fujistsu indicates 100 mA current and 5 V for its miniature relays with contacts made of popular alloy AgNi for general commercial use; Kuhnke declares 50 mA and 20 V for the same type of contacts and Phoenix specifies 100 mA and 12 V.

In many cases, these minimum rates of current and voltage exceed real rates in circuits of operational amplifiers, logic elements, and other low-rate components of modern electronic circuits. If we consider that these rates are valid for the date of relay delivery and will increase (due to deterioration of conductivity of contacts) during the next 5–10 years of BC use, it becomes obvious why this problem is so important.

Contacts with coatings of gold are used for switching of low-rate and "dry" circuits. The same companies produce such relays in large numbers. What should we do if we start experiencing low conductivity of contacts made of silver-based alloys in a specific BC? Should we take them out of the plates and substitute new contacts? This is possible, if one can find an absolutely suitable relay in terms of parameters and location of its pins. In some cases (not in all), one can try to open a relay and clean the contacts with a special tool.

We suggest using another solution: the self-cleaning of a relay's contacts with electric spark. For this we use a very simple, low-frequency multivibrator based on a 555 series chip with a small relay at the output (Figure 3.32).

FIGURE 3.32
Low-frequency multivibrator based on 555 series microchip and a miniature output relay.

The contacts of the output relay of this multivibrator close and open at 0.5 Hz frequency. They are connected to the supply line of external relay winding—the contacts of which are cleaned. These contacts, in turn, are connected to a 220 V AC line in series with active load (resistance) for 1–2 A current (incandescent lamps should not be used as a load due to high rates of start inrush current). Self-cleaning of the relay's terminals by a moderate spark with 0.5 Hz lasts 15–20 minutes. During this time, oxide and sulfide films burn in the contact area and the transition resistance of contacts returns to its normal rating.

3.11 A Device to Monitor Proper Ventilator Functionality in Sites with Batteries

According to the electrical installation code, premises where batteries are installed should be equipped with forced ventilation systems. Moreover, the code contains requirements to the airspeed in the rooms with batteries installed when the ventilation system is working. Also, the code suggests that there should be a block system, which would not allow the battery's charge with a voltage exceeding 2.3 V to a cell when ventilation is off.

Unfortunately, these requirements are not enough to ensure safe work in rooms with batteries (these rooms may be adjacent to production premises when necessary). For example, the airspeed in the ventilated rooms is only a calculated (not a controlled) parameter, and the block system mentioned in the electrical installation code is based on control of relays' contacts' position, which turns a ventilator on or off and cannot react to malfunctions of the ventilator. The malfunction may include burn of the ventilator motor winding; contact failure or breaking its circuit; dirty ventilator and mechanic overload of the ventilator's motor, which results in reduction of ventilator capacity; short-circuit in the supply line; power loss; etc.

Since there are no common solutions to these problems, we have developed a new device to monitor the ventilator's condition. It should be mentioned here that the most efficient and correct technical solution would be to monitor the air flow in the ventilation system constantly. However, considering the practical difficulty of such monitoring implementation, we offer a simple, but rather efficient, device based on ordinary relays that can monitor conditions of a ventilator by using electric parameters of its motor. We can say that this option of monitoring is optimal in terms of cost efficiency.

The device (Figure 3.33) consists of only three relays and a protection circuit breaker, located in a small plastic housing. The main tool of this device is represented by a special current relay K1, which supervises the range (window) of current rates. It is necessary to set up lower and upper limits of

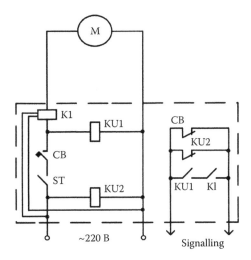

FIGURE 3.33
Electric circuit of the device to monitor the ventilator's condition in the rooms with batteries installed.

current rates under supervision and time delay. When the supervised current rate goes beyond these limits, the relay is activated. The time delay is necessary for detuning from the start inrush current of motor.

Relays to supervise the current rate range (window) are manufactured by different companies (Figure 3.34). One can also use a simple programmable controller with a miniature transformer of current, but this option is more expensive and more difficult than the use of a specialized relay.

In addition to this relay, this device includes relays for voltage control, KU1 and KU2, which are realized by the ordinary auxiliary electromagnetic relays with a coil for 220 V AC as well as a CB with a special auxiliary contact. The special point of this auxiliary contact is that it is not activated during manual manipulations with a CB, but rather is activated (closes) when the CB release is activated—that is, during automatic activation of a CB under overload or short circuit. Such CBs and auxiliary contact to them are manufactured by different companies. For example, a switch FAZ-D1/1-RT and its auxiliary contact Z-NHK are manufactured by Moeller.

Occasional automatic on/off switching of a ventilator is achieved by means of a programming timer ST. The principle of operation is as follows:

> When the ventilator and its motor are in good shape, the current consumed by the motor falls within a narrow boundary window range of rates, which are supervised by relay K1; its contact is open. KU1 and KU2 relays are in a closed position under the voltage applied to them. KU1's contact is closed and KU2's contact is open. Auxiliary contact of the CB is open. A general alarm circuit is open.

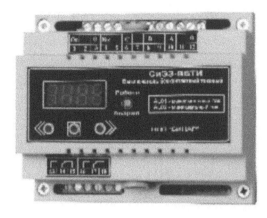

CM-SFS.22   SiЭZ-VBTI-0.5-60
(ABB)   (NPP "Binar")

FIGURE 3.34
Relay for supervision of current rate range ("current window"), manufactured by ABB and a Belarus company NPP "Binar" for current rates from 0.3 to 15 A.

When a ventilator becomes dirty or in case of increased friction in its bearings, the mechanical load on the ventilator's shaft is increased; thus, the motor consumed current also increases. When the current rate exceeds a preset range, the K1 relay will be activated, closing output contact K1, which is connected in series with the already closed KU1 contact. The general alarm circuit will close.

In case of motor burns, contact failure in the supply line, or breakdown in wiring, the current flowing through the K1 relay will fall lower than the present limit and a relay will be activated, closing its contact in series with the already closed KU1 terminal. The general alarm circuit will close.

In the event of a short circuit in the supply line of the motor, a protection switch CB will be activated disconnecting the supply line. Its a-switch will close a general alarm circuit regardless of the condition of other elements of the circuit.

In case of power loss in the main supply, a coil of the KU2 relay will be deenergized and the KU2 contacts will close a general alarm circuit regardless of the position of other elements of the circuit.

In case of breakdown of supply line of the motor by a preset timer ST, a coil of the KU1 relay will be deenergized and its KU1 contact will open, opening a general alarm circuit even though the K1 contact will be closed at this time.

If ventilation is disconnected manually by means of the CB, the K1 terminal will close and the KU1 terminal will open. The auxiliary CB contact will stay in its initial (open) condition. The general alarm circuit will remain open.

An inrush of start current during motor switching-on does not result in the closing of the alarm circuit since there is time delay in the K1 relay. Thus, a simple device can solve a problem of monitoring a ventilator's condition at low cost in rooms with batteries, preventing formation of dangerous concentrations of hydrogen and acid evaporation in the air.

References

1. *A manual on designing auxiliary DC power supply systems*—PS ENES. STO 56947007-29.120.40.093-2011.
2. Bogdanov, D. I. 1972. *Ferroresonant voltage stabilizers*. Moscow: Energiya.
3. Gurevich, V. 2006. *Electric relays: Principles and application*. Boca Raton, FL: CRC Press (Taylor & Francis Group).

4
Uninterruptable Power Supply

4.1 Diversity of Uninterruptable Power Supplies

If the battery chargers (BCs) described in a previous chapter represent an important component of auxiliary DC power supply systems, uninterruptable power supplies (UPSs) are no less important for auxiliary AC power supply systems and in general for systems of 0.4 kV substations and power plants.

There are different varieties of UPSs according to type, principle of operation, and design. According to international standard IEC 62040-3 [1] there are three main types of UPSs:

- Passive standby, which was formerly called off-line (IEC 62040-3.2.20)
- Line interactive (IEC 62040-3.2.18)
- Double conversion, which was formerly called online (IEC 62040-3.2.16)

The passive standby UPS (Figure 4.1) is the simplest and cheapest. Under normal operation, the input receives power directly from the main supply line. When power is lost in the main supply line, the device switches to an inverter and a battery by means of a quick switch (within 3–10 ms). This means that when using such a UPS, there can be short-term interruption in power supply ranging from several milliseconds to half of the period of supply voltage. This is one of the drawbacks, though these interruptions are not critical for most operations. Another drawback is that in the normal operation mode all unfavorable effects come into the UPS from the main supply line (e.g., distortion of voltage form, impulse overload, deviation of frequency and voltage, and others). To compensate for these drawbacks, the passive standby UPS is equipped with additional passive filters of harmonics, varistors, etc. As a rule, this principle is employed for the cheapest and least powerful UPSs (up to 1.5–2 kW)—for example, household UPSs for computers (their power supply allows short-term voltage interruption).

A line-interactive UPS (Figure 4.2) does not have a separate charger for the battery. It receives power from the output of a continuously operating rectifier converter. The input is continuously received power from the main

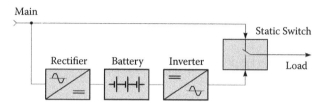

FIGURE 4.1
A structure of passive standby UPS.

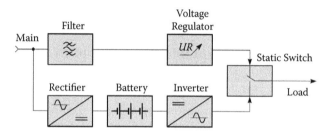

FIGURE 4.2
Design of line-interactive UPS.

supply line through a quick switch, similar to that of passive standby UPSs. The main distinction of a line-interactive UPS is that it has a control device (stabilization) for output voltage *UR,* which is controlled by a microprocessor that monitors main line voltage and reacts to its change. This is actually why the UPS is called line interactive (another name: voltage independent). The voltage controller stabilizer can be electronic or ferroresonant (in a UPS of relatively low power). It can also be made as part of an automatic tap change transformer or boost transformer, the primary winding of which is connected in series to the main power line, while the secondary winding—energizing winding—is connected to an additional inverter with a power 20% that of the main inverter. In the latter case, the UPS is called UPS with delta conversion. This technology was developed and patented by Silicon Group, which is now a division of American Power Conversion.

If the voltage in the main supply line is lost at all or exceeds the preset, comparatively wide range in which the voltage is controlled, the power is switched to the supply from the inverter and a battery after activating a quick switch. In any case, galvanic decoupling of the load from the main supply line in this type of UPS is not provided and voltage frequency in the device is determined by the supply line.

The line-interactive UPS has a power from 1 to 10 kW and is widely used in offices and for power supplies to commercial computer equipment of limited power.

The most advanced and most expensive UPSs are those with double conversion, where input AC voltage of the main line is initially converted into DC

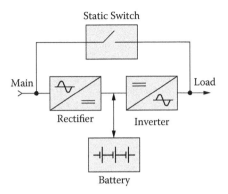

FIGURE 4.3
UPS with double conversion.

voltage and then this DC voltage is converted back to AC again (Figure 4.3). The output of AC voltage of such a UPS does not depend on main supply line voltage in terms of both level and frequency, since it is produced from DC voltage at the output of a rectifier by a continuously operating inverter under any regime. This is the reason the UPS is called voltage-frequency independent. This UPS is implemented for power supply of more important consumers (e.g., large corporate servers and control systems of uninterruptable processes, as well as for building a centralized network of power supply) that require complete insulation from the supply line.

4.2 Static Switch

A quick switch in the UPS with double conversion does not participate in the normal operation of the UPS, but is used only in case of inverter overload or any other problem with the UPS. In this case, a consumer is automatically, quickly switched directly to a supply line where a phase angle of alternating current in the circuit load remains unchanged. If switching to a bypass was caused by overload in the external line, after elimination of the problem, the UPS is automatically returned from bypass to normal operation mode. Such a switch is called a bypass switch. Automatic bypass is not installed in all the double conversion UPSs; for example, it does not make sense in the lower capacity of the passive standby UPS.

To ensure quick activation of a bypass switch, it is constructed based on a powerful semiconductor switch, often on two back-to-back thyristors in each input (Figure 4.4) and that is why it is called a static switch. Lengthy flowing of full load current through a static switch will cause it to emit a large amount of heat. Therefore, thyristors should be (1) calculated for a full load

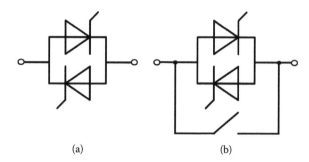

(a)   (b)

FIGURE 4.4
A simplified layout of a power part of a static switch for one phase.

current with corresponding reserves, and (2) equipped with large and heavy radiators. This can be avoided if, after activation of a static switch, thyristors will be quickly shunted by terminals of an ordinary electromagnetic contactor (Figure 4.4b).

However, it is thought [2] that the reliability of the static switch in Figure 4.4(b) is lower than the reliability of the switch in Figure 4.4(a) with thyristors selected with corresponding reserves in terms of current and voltage.

Static switches are manufactured not only as an internal part of a UPS, but also as separate products (Figure 4.5), which are used for quick and synchronized supply line switching of consumers from the main power supply to a backup, switching between different UPS types, etc.

In addition to automatic bypass, almost all commercial UPSs of medium or large capacity have a manual bypass designed to allow servicing of the UPS without delays in power supply. It is manufactured as part of a usual switch

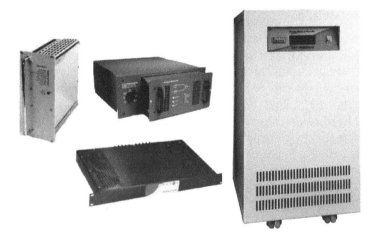

FIGURE 4.5
Static switches of different capacities manufactured as separate products.

of sufficient capacity, with a manual drive connecting the load circuit with the supply line directly, bypassing the UPS.

4.3 Inverter

DC to AC inverters are an important and integral part of all UPS types. As a rule, power inverters are based on powerful field effect transistors (FETs) or *insulated gate bipolar transistors* (IGBTs), working in a switching mode—in other words, opening and closing with a frequency of tens to hundreds of kilohertz.

Inverters based on a half-bridge layout (Figure 4.6a) or a full bridge layout (Figure 4.6b) are used more often. In the half-bridge layout (Figure 4.6a) capacitors C1 and C2 are charged to half of the power source voltage each ($U_{C1,C2} = 0.5E$). When transistor 1 is opened, capacitor C1 is discharged through the primary winding of transformer T. This pulse of discharging current flows through primary winding of a transformer and creates voltage in the secondary winding. When transistor 1 is closed and transistor 2 is opened, the current of capacitor C2 is discharged through the same winding of the transformer, but in the reverse direction. Thus, pulses of current with

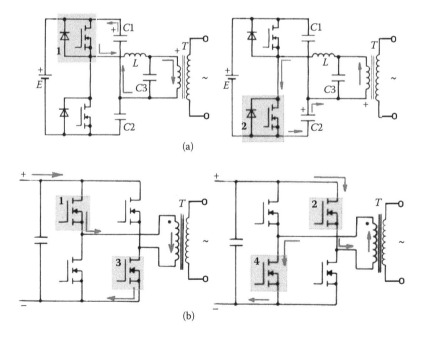

FIGURE 4.6
Principle of conversion of direct voltage of a battery into alternating voltage by an inverter in a UPS.

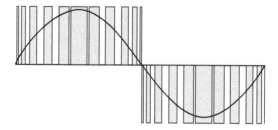

FIGURE 4.7
Principle of pulse-width modulation (PWM).

opposite polarity (i.e., alternating current) flow through primary winding of transformer T.

Strictly speaking, transistors do not create only one pulse; they create a group of pulses of variable width. The control of the rate of output voltage is performed by changing the width of the pulses that control transistors (i.e., by changing the time during which transistors are working in the conducting state). This principle of transistor management is called pulse-width modulation (PWM; Figure 4.7).

Due to the transformer's inductivity, groups of pulses of opposite polarity current (created by transistors) start resembling ordinary sinusoids in form. Alternating voltage created in the secondary winding of the transformer is delivered to the customer.

In the full bridge layout (Figure 4.6b), two additional transistors are used instead of two capacitors. In this layout, a series of sequential openings and closings of transistor pairs (1 and 3, and then 2 and 4) result in the creation of bipolar groups of current pulses in the primary winding of transformer T. The rest in this layout is similar to that of the half-bridge layout. As a rule, there are additional smoothing filters at the inverter's terminals, which include powerful chokes and capacitors.

In practice, UPS types are not always used in their classical design, as we discussed previously, because independent inverters and static switches are available on the market. They sometimes use an inverter and a static switch in addition to what is already installed at most substations and power plants, a battery, and a battery charger to obtain a fully operational UPS.

4.4 Group Powering of UPS

Group powering of the UPS is employed when the capacity of one UPS is not sufficient, as well as when it is necessary to ensure reliable power supply of a consumer. In the simplest case, parallel powering of two UPSs is used. Such

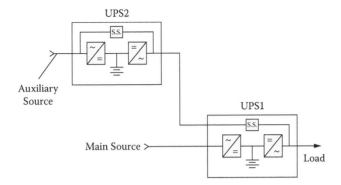

FIGURE 4.8
A layout of powering two UPS in series.

connection is sometime called "parallel redundant UPS configuration." This can only be done with UPSs of similar type and equipped with a special unit for load balancing between the two UPSs (load sharing controller). Another thing is powering of two UPSs in series (sometime named "isolated redundant UPS configuration") (Figure 4.8).

These can be different types of UPSs without any additional units for load balancing. With this configuration, UPS 1 is operating continuously, receiving power from the main power supply, while UPS 2 is in so-called "cold reserve." If UPS 1 fails, its static switch (SS) turns UPS 2 on, which receives power from an auxiliary power source (or from the same source). This connection is used to increase reliability of the power supply to consumers and it is thought that it provides a higher rate of reliability than a parallel connection.

Two separate AC power supplies are used to increase reliability of a single UPS as well (Figure 4.9).

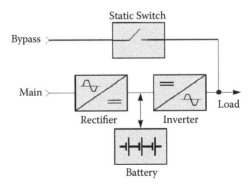

FIGURE 4.9
UPS with double conversion, connected to two power supply lines.

4.5 A Problem of Energy Quality in Networks with UPS

With the increase in capacity of the modern UPS (it can reach as much as 100 kW and even several megawatts; Figure 4.10), a problem of "pollution" of a network with high-frequency harmonic waves, which results from the distortion of the current form consumed by a powerful UPS, has become acute. These UPSs, as previously mentioned, perform double conversion and provide a very good quality of energy to consumers connected to their output, but at the same time they "pollute" the supply line, from which they consume energy.

Any distorted (or nonsinusoidal) periodic signal (current, voltage) can be represented mathematically as one that includes a set of clear sinusoidal signals, among which a signal of main frequency and a set of sinusoidal signals with a frequency divisible by the frequency of the main signal can be selected. For example, if the frequency of the main signal (fundamental frequency) is 50 Hz, then the frequency of the second harmonic will be 100 Hz and the frequency of the fifth harmonic will be 250 Hz. The signals with frequency higher than the main frequency are called *harmonics* and the signals with frequencies lower than the main frequency are called *subharmonics*. The splitting of a nonsinusoidal (distorted) periodic signal into a series of sinusoidal signals is called *Fourier series decomposition*. The idea of this method is based on the fact that it is always possible to select a series of harmonic (i.e., sinusoidal) signals with amplitudes, frequencies, and initial phases such that their algebraic sum will be equal to the ordinate of a resulting nonsinusoidal signal at any time (Figure 4.11).

FIGURE 4.10
Heavy-duty UPS.

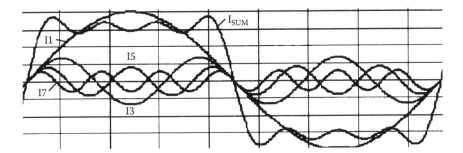

FIGURE 4.11
Resulting nonsinusoidal current (I_{sum}) and its forming sinusoidal currents of the fundamental frequency (I_1), as well as those of the third (I_3), fifth (I_5), and seventh (I_7) harmonics.

The presence of harmonics in electric lines negatively influences both the lines (increase of current in the neutral conductor, additional losses and heating of wires and cables, premature aging of their insulation, resonance events) and almost all types of electric equipment, except for the simplest heating units. Harmful leakages appear both in telecommunication and control lines, increased sonic noise appears in electromagnetic equipment, vibrations in electric machine systems increase and insulation of electric machines dries out at accelerated speed, and transformers and capacitors get overheated. These are the reasons the accepted level of harmonics in electric networks is limited by standards.

The number of harmonics in electric networks is represented by a coefficient called total harmonic distortion (THD), which is the ratio of the root-sum-square value of the harmonic content to the root-mean-square (RMS) value of the fundamental (first) harmonic. For voltage, this coefficient is calculated according to the following formula:

$$THD_\% = \frac{\sqrt{V_2^2 + V_3^2 + V_4^2 + \ldots + V_\infty^2}}{V_1} \times 100\% \qquad (4.1)$$

where $V_2, V_3, V_4, \ldots$ are voltages that correspond to the second, third, fourth, …harmonics and V_1 is the first (fundamental) harmonic's voltage

There is a sort of misunderstanding in the Russian technical literature as to what term should be used—"coefficient of nonlinear distortions" (CNDs) or "total harmonic distortions" (THDs)—because these different definitions are often mixed up. They have been used interchangeably in order to define what State Standard 13109-97 [3] calls a "coefficient of distortion of sinusoidal voltage curve." We will use the latter as an equivalent of the International Standard THD.

The formula to calculate the THD by current is the same except that it uses current harmonics instead of voltage harmonics.

Due to a significant negative influence of harmonics on electric equipment, the maximum level of the THD based on voltage is limited by the International IEC Standard and the American IEEE Standard at 5% for general purpose electric lines with voltage ranging from 120 V to 69 kV. In some other standards (e.g., British standards G5/4-1:2005 [4] and BS EN 50160:2010 [5] as well as in Russian State Standard 13109-97 [3]), the acceptable voltage-based THD rate amounts to 8% for lines of 0.4 kV.

As for the limits of current-based harmonic distortions, everything is much more difficult here and standards do not provide a specific rate that could be applied for any and all situations, since the maximum rate of current-based harmonic distortions in the electric networks depends on the network itself—particularly on impedance, which is determined by the maximum current rate of network short-circuiting. The Russian State Standard 13109-97 does not provide any limits for current distortions at all. The dependence of harmonics' permissible level in a network on its impedance can be illustrated with a sample part of a circuit with a powerful nonlinear load, connected at the end of a long-haul line and relatively low-power linear loads distributed in the line (Figure 4.12).

A 100 A nonsinusoidal current with a THD of 25.5% flows through a long-haul (for example, cables) line creates voltage drops in sections of this line. The higher the impedance of the line is, the larger is the voltage drop of each harmonic created by a corresponding harmonic of current. As a result, a voltage source for low-power linear loads will be represented by nonlinear voltage, which determines the flow of nonlinear current through linear load with all possible negative consequences. This explains why standards match

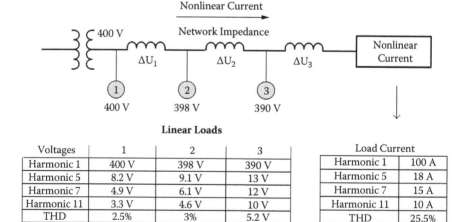

Voltages	1	2	3
Harmonic 1	400 V	398 V	390 V
Harmonic 5	8.2 V	9.1 V	13 V
Harmonic 7	4.9 V	6.1 V	12 V
Harmonic 11	3.3 V	4.6 V	10 V
THD	2.5%	3%	5.2 V

Load Current	
Harmonic 1	100 A
Harmonic 5	18 A
Harmonic 7	15 A
Harmonic 11	10 A
THD	25.5%

FIGURE 4.12
Distribution of current and voltage harmonics in a network with a powerful nonlinear load and linear loads distributed in the line.

TABLE 4.1

Limiting Values of Current Distortions in General Purpose Lines
with Nominal Voltage Rate from 120 V to 69 kV

I_{SC}/I_L	h < 11	11 ≤ h < 17	17 ≤ h < 23	23 ≤ h < 35	35 ≤ h	TDD
<20	4.0	2.0	1.5	0.6	0.3	5.0
20–50	7.0	3.5	2.5	1.0	0.5	8.0
50–100	10.0	4.5	4.0	1.5	0.7	12.0
100–1000	12.0	5.5	5.0	2.0	1.0	15.0
>1000	15.0	7.0	6.0	2.5	1.4	20.0

Source: IEEE Standard. 519-1992. IEEE recommended practices and requirements for harmonic control in electric power systems. IEEE Industry
Applications Society/Power Engineering Society.

Note: I_{SC}—short-circuit current; I_L—maximum load current; h—number of
harmonics; TDD—total demand distortion.

maximum permissible rate of current-based distortions with parameters of
the line, since the same nonlinear current will differently affect consumers connected to the line depending on the impedance of the line's sections
(Table 4.1).

Differences of current THD from total demand distortion (TDD) are

$$I_{THD} = \frac{\sqrt{I_2^2 + I_3^2 + I_4^2 + I_5^2 + \ldots}}{I_1} \times 100\% \qquad (4.2)$$

$$I_{TDD} = \frac{\sqrt{I_2^2 + I_3^2 + I_4^2 + I_5^2 + \ldots}}{I_L} \times 100\% \qquad (4.3)$$

In other words, if a current's THD is a ratio of RMS voltage of the sum
of higher harmonics of current, except the first, to the first harmonic current (I_1), then TDD is a ratio of RMS voltage of the sum of higher harmonics of current, except the first, to the maximum load current (I_L). The IEEE
Standard 519 [6] recommends using the average rate for 12 months as a
maximum load current. Under full nominal load, $I_{THD} = I_{TDD}$ (Table 4.2).

It is obvious from Table 4.2 [7] that total amount of line "pollutants" (I_{THD})
is increasing with the reduction of electric drive load (due to the inverter's
mode of work), while the I_{TDD} estimate is decreasing, which means reduction of harmful influence of electric drive on the line. Thus, very large coefficients of current harmonic distortions in the circuit of nonlinear load may
not create significant influences on other consumers connected to the same
line—particularly if the power of the line (maximum short-circuit current)
is significantly higher than the power load (large estimate of I_{SC}/I_L).

TABLE 4.2

Current Rates Measured in a Circuit of a Six-Pulse, Variable-Speed Electric Drive with a Power Rating of about 150 kW

	Current, A		Current Harmonics, A					
Load	RMS	Fundamental	5th	7th	11th	13th	I_{THD}	I_{TDD}
Full	233	182	118	80	12	12	79%	79%
75%	187	142	96	70	15	7	86%	65%
50%	134	96	69	54	17	5	96%	48%
25%	67	43	33	29	14	9	120%	30%

Source: Hoevenaars T., K. LeDoux, and M. Colosino. 2003. IEEE Industry Applications Society 50th Annual Petroleum and Chemical Industry Conference, 2003. Record of conference papers, Sept. 15–17.

TABLE 4.3

Norms of Standards of Harmonic Current Parts for Symmetric Three-Phase Loads

Nominal value, R_{SCE}	Max. Permissible Current Harmonic Limits, I_n/I_1, %				Max. Permissible Limits for Coefficients	
	I_5	I_7	I_{11}	I_{13}	ACHP	PWCHP
33	10.7	7.2	3.1	2	13	22
66	14	9	5	3	16	25
120	19	12	7	4	22	28
250	31	20	12	7	37	38
≥350	40	25	15	10	48	46

Source: Russian State Standard P 51317.3.12–2006.
Note: I_n—value of n-harmonic; I_1—value of fundamental component.

Using a methodology offered by the IEEE Standard 519, it is possible to evaluate the influence of current distortions.

Russian State Standards P 51317.3.2-2006 [8], P 51317.3.5-2006 [9], and P 51317.3.12-2006 [10] offer a little bit different approach, providing reference charts of maximum permissive current rates (percentage in relation to the first harmonic) depending on load current (<16 A, >16 A, <75 A) for each harmonic (Table 4.3) as well as generalized indicators for harmonics.

$$ACHP = \sqrt{\sum_{n=2}^{40}\left(\frac{I_n}{I_1}\right)^2}, \quad PWCHP = \sqrt{\sum_{n=14}^{40}\left(\frac{I_n}{I_1}\right)^2} \tag{4.4}$$

where ACHP is aggregate coefficient of harmonic parts and PWCHP is partial weighted coefficient of harmonic parts:

$$R_{SCE} = \frac{U_{NOM}}{\sqrt{3} \cdot ZI_{equ}} \qquad (4.5)$$

where I_{equ} is maximum root-mean-square rate of current.

In addition to indicators of current and voltage harmonic compositions mentioned previously, there are other indicators, such as crest factor, which is sometimes called a coefficient of amplitude (Ka), or peak factor. The crest factor is a relation of amplitude rating of a signal to its RMS (Figure 4.13). In addition to the crest factor, there is also a form factor, which is a ratio of RMS of a signal to its average rating. For sinusoidal signal, the crest factor is 1.41 and the form factor is 1.11.

Another indicator, which describes the influence of current distortion (voltage), that is created by nonlinear loads in electric equipment is the K-factor. K-factor is a coefficient of increase of losses in a power transformer due to nonlinearity of load. According to standards of Underwriters Laboratories (UL) UL1561 [11] and UL1562 [12], K-factor is calculated by the following formula:

$$K = \sum_{h=1}^{\infty} (I_{h(pu)})^2 h^2 \qquad (4.6)$$

where $I_{h(pu)}$ is root-mean-square rating of harmonic current h consumed by the load and h is number of harmonics.

You can come across this formula in the technical literature with a reference to IEEE Standard C57.110/D7-1998 [13]. This is a common mistake, since

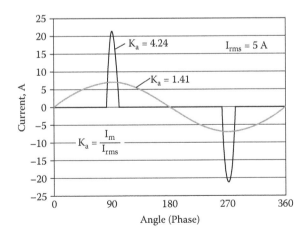

FIGURE 4.13
Crest factor (coefficient of amplitude—K_a) for sinusoidal and nonsinusoidal current with similar RMS rating (I_{rms}) 5 A.

this standard offers another coefficient of influence of nonsinusoidal current on power transformers and is called the harmonic loss factor (F_{HL}):

$$F_{HL} = \frac{\sum\limits_{h=1}^{h=h_{max}} I_h^2 h^2}{\sum\limits_{h=1}^{h=h_{max}} I_h^2} = \frac{\sum\limits_{h=1}^{h=h_{max}} \left[\frac{I_h}{I_1}\right]^2 h^2}{\sum\limits_{h=1}^{h=h_{max}} \left[\frac{I_h}{I_1}\right]^2} \tag{4.7}$$

There is a relation between these two coefficients:

$$K = \left[\frac{\sum\limits_{h-1}^{h=h_{max}} I_h^2}{I_R^2}\right] F_{HL} \tag{4.8}$$

There is no direct correlation between THD on the one hand and the K-factor or F_{HL} on the other hand; this is why IEEE Standard C57.110 provides tables with the results of calculation of F_{HL} for different sets of harmonics in the current load. There are also tables with calculations of K-factor for different types of loads. Unfortunately, if we know the K-factors of separate individual loads, we will not be able to determine its total rating, since the increase of heterogeneous harmonics in the line, produced by different loads, results in a decrease rather than increase of "pollution" of the line by harmonics and consequently to a decrease of both THD and K-factor. This happens as a result of compensating influence of harmonics created by different nonlinear loads on each other. However, we can get useful information from these charts. You can see from the charts that for a load represented by incandescent lamps K = 1, for UPS without a filter at the input K = 13, and for the same UPS but with a filter K = 4.

A fully loaded dry transformer at K-factor = 4 emits 10% more heat than under linear load (K = 1), while at K = 13, the transformer emits 25% more heat, which should be diverted somewhere if the transformer is fully loaded; otherwise, the load of the transformer should be reduced in the same amount (Figure 4.14). Particularly, Figure 4.14 [14] shows that if there is a powerful UPS without a special filter as a load (K = 13) of a power transformer at a substation or a power plant for its own needs, the power of such a transformer should be at least 25% more than the power of a load. This increase amounts to about 10% for a UPS with a filter (K = 4).

To supply powerful nonlinear loads, special transformers with improved cooling systems are manufactured, for which a maximum K-factor of a load

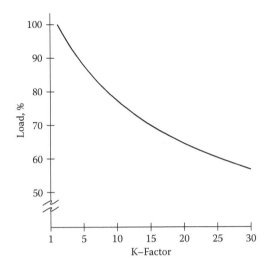

FIGURE 4.14
Dependence of maximum permissible load of a power transformer on K-factor of its load. (Deshpande, K., R. Holmukhe, and Y. Angal. 2011. K-factor transformers and non-linear loads. National Conference 2011, Bharati Vidyapeeth Deemed University, College of Engineering.)

is indicated. At this rated K-factor, the transformer can be 100% loaded (usually K = 1 for a common transformer).

Now it becomes clear why the problem of current and voltage distortion brought into a supply network by powerful nonlinear loads is so important and why we should control them. A UPS represents such a nonlinear load. Which measures to control these distortions are implemented in the network with a powerful UPS?

Six-pulse rectifiers implemented in a three-phase UPS create a high level of the fifth current harmonic in the supply network (Figure 4.15). In order to reduce distortions, 12-pulse layouts are implemented in the UPS that consist of two transformers with secondary windings connected in a delta and star layouts and two rectifying bridges, the outputs of which are connected in parallel (Figure 4.15).

Another way to reduce the UPS's influence on the supply line is the application of powerful resonant filters of the fifth harmonic (the most significant in the spectrum), which include chokes and capacitors (Figure 4.16).

Another efficient measure to reduce the influence of the UPS on the supply line is to reduce the impedance of the distribution line (see Figure 4.12). This is achieved by increasing the section of cables (wires). A maximum efficient section of cable (wire) cores amounts to 95 mm², roughly. With further increases of cable sections, their inductivity remains relatively constant, so in order to further reduce impedance of cables, they are connected in parallel. Reduction of the influence on the supply line can be

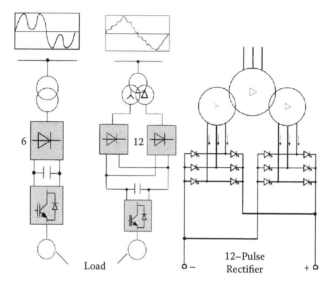

FIGURE 4.15
Simplified flow diagram of a UPS with 6-pulse and 12-pulse circuits of input rectifier, current that they consume from the line, as well as a detailed diagram of a 12-pulse, three-phase rectifier.

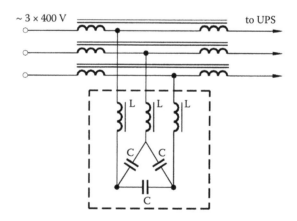

FIGURE 4.16
Passive resonant filter of the fifth harmonic for UPS.

achieved by splitting the line into fragments using a less powerful UPS for each of them.

A rather efficient way of controlling harmonics in the lines with a powerful UPS is application of active harmonic filters (AHFs). They analyze harmonic composition of current in the circuit with a nonlinear load and generate harmonics into the line, which are in the opposite phase to load current harmonics (Figure 4.17).

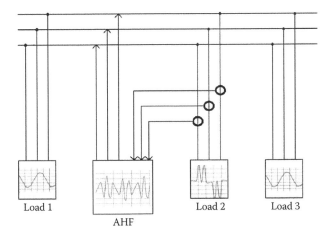

FIGURE 4.17
A diagram of connection of an active harmonic filter (AHF) into the line with powerful source of harmonics (load 2).

The two main types of AHFs differ in the way in which they are connected into the line: in series and in parallel. AHFs connected in series are introduced into the electrical open by means of boost transformers. The winding with a low number of coils wound as a thick cable is connected in series to the input of the supply line and the winding with a bigger number of coils, and a thin cable is connected to the output line of the AHF. These devices are pretty large and expensive. AHFs with parallel connections enjoy more widespread use (Figure 4.17).

Application of an AHF proves to be very efficient (Figure 4.18), unlike passive resonant filters, including noncharacteristic and low-frequency ones. These devices are manufactured today by many companies and they are available on the market in a whole range of powers for currents from tens to hundreds of amperes (Figure 4.19).

4.6 High-Voltage UPS Aggregate

Equipment used in modern industrial plants (process controllers, programmable logic controllers, adjustable speed drives, robotics) is actually becoming more sensitive to voltage sags as the complexity of the equipment increases. Even relays and contactors in motor starters can be sensitive to voltage sags, resulting in shutdown of a process when they drop out. The semiconductor-producing factories require especially high levels of power quality due to the sensitivity of equipment and process controls. During certain stages, light is applied to photosensitive layers to create the circuit paths. A microprocessor

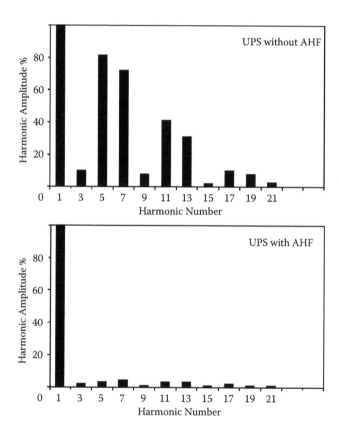

FIGURE 4.18
Spectral composition of current, consumed by UPS without a filter from the line, and current
in the long-haul line with activated AHF.

FIGURE 4.19
Layout of AHF manufactured by different companies.

FIGURE 4.20
Separate UPS units included inverter, 0.4 kV battery rack, and step-up transformer 0.4/22 kV.

is made up of hundreds of such layers. If a significant voltage sag occurs during the typically 7- to 15-day production cycle, the entire work in process could be scrapped. In this instance, costs for one incident could total $200,000 to $2,000,000. As semiconductor-processing equipment is especially vulnerable to deep voltage sags and interruptions, the SEMI F47 standard defines the voltage sag ride-through capability required for semiconductor processing, metrology, and automated test equipment. High-voltage UPS aggregates are often used for power supply of whole semiconductor factories. Such UPSs include several separate units (Figure 4.20) and a high-voltage static switch incorporated in high-voltage switchgear (Figure 4.21).

Each UPS unit includes an inverter with battery rack for 400 V and step-up power transformer of 0.4/22 kV. All UPS units connect in parallel on a 22 kV bus bar in switchgear. Such aggregates with four UPS units (Figures 4.20 and 4.21) work as single, high-voltage passive standby UPSs with 10 MVA power.

4.7 Additional Problems of the UPS

The same system supplies driving gears of power switches and many other devices, causing spikes and surges. The organizations of the supply system for relay protection by a UPS cushion the negative impacts of such factors. Many specialists are sure that a UPS provides full solution for such problems.

FIGURE 4.21
High-voltage (22 kV) thyristor-based static switch with bus bar incorporated in special HV switchgear.

However, investigations of UPS systems [15] have shown that, at certain conditions, noise spikes and high harmonics can get into load (microprocessor-based relay protection devices and other sensitive electronic devices) through grounded circuits and that neither a UPS nor filters can prevent this.

In addition, UPS devices have their own changeover times. Usually, specifications for UPSs indicate a switch delay time of 3–5 ms, but in fact under certain conditions this time may increase by a factor of more than 10. In a normal mode, the load in some types of UPS devices ("passive standby" types) is usually fed through a thyristor switch (static switch), which must become enabled when voltage diverts below 190 or above 240 V. After that, the load is switched to the output of the inverter supplied from the accumulator battery. The time of disabling of the thyristor switch is summed up from the time of decrease of current flowing through the thyristor to the zero value t_0 and turn-OFF time t_q, after which the thyristor is capable of withstanding the voltage in the closed (OFF) position.

For different types of thyristors, $t_q = 30–500$ μs, and t_0 depends on the proportion of induction and pure resistance of the circuit of the current flowing through the disabled thyristor (if the circuit breaks). As a result of switching-OFF of the input circuit breaker or pulling out of the UPS supply cable from the terminal block, $t_0 = 0–5$ ms, which will provide very high performance; however, at actual interruption of supply, a break usually occurs in circuits of a higher voltage level, but not in the circuits for 120, 220, 380, or 400 V. The thyristor switch is then shorted to the second winding of the network transformer and the load connected to it. If the beginning of the supply interruption coincides with the conductivity interval of the thyristor, the duration of the process of current decaying may exceed 400 ms, and changeover to reserve power supply (inverter) may be delayed for inadmissibly long times. If the beginning of supply interruption coincides with the no-current interval, switching requirements are similar to those at input cable break.

It follows that at interruptions of supply and voltage fall-throughs coinciding with the interval of current flowing, the working source will change over to the reserve supply for inadmissibly long times [16]. The user who sets up and checks the UPS usually reaches conclusions about its serviceability by switching the input circuit breaker OFF. As shown earlier, this does not always correspond to conditions of actual transient processes (short-circuiting of the input to the induction shunt). This creates the possibility that, after installation and successful check of the UPS by switching OFF the input circuit breaker, some voltage fall-throughs (coinciding with intervals of current flowing) may lead to disabling ("suspension") of the microprocessors. That is why, in some types of UPSs of the passive standby category, in order to provide high performance during changeover in the short-circuit mode, a high-speed electromechanical relay instead of thyristor ones is used as a static switch.

To determine the actual time of changeover of the UPS, one should undertake an experiment imitating short circuits of the working source to activate a shunt inductively while a special measuring system records the transient processes. But who runs such experiments and where?

Thus, another aspect of the problem gains our attention: suspensions and malfunctions of operation of the microprocessor of the UPS in emergency modes on high-voltage circuits. When the control microprocessor malfunctions, alternation of switching ON and switching OFF power semiconductor elements of the inventor and short-circuit loop making may be disturbed, followed by automatic switching OFF the input circuit breaker of the UPS. This same phenomenon can happen to automatic chargers whose microprocessors are supplied from an external auxiliary UPS. Such incidents quite often occur in practice, but nobody yet has attempted a serious analysis of the reasons. It is quite possible that the reasons for such emergency switching of USPs and of the chargers are similar to those for the case considered previously.

4.8 Dynamic and Hybrid UPSs

In concluding this chapter we should mention the dynamic rotary UPS (Figure 4.22). This device uses an electric motor that constantly turns a power generator, feeding the load. There is a flywheel on the shaft that they share. A

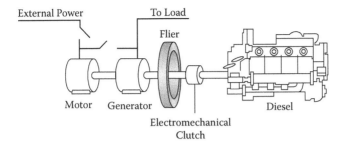

FIGURE 4.22
Principle of operation of a dynamic rotary UPS.

shaft of a diesel engine is connected to this shaft through an electromechanical clutch. In case of power loss, the generator continues turning and produces power within time frames necessary to start the diesel engine. When the electric motor starts receiving power from the outside source, the diesel is shut down. Thus, an uninterruptable power supply to consumers is achieved.

Another type of UPS that has become increasingly popular is a hybrid device based on a super-flywheel with a generator and inverter (Figure 4.23). In this device, a super-flywheel with bearings on magnetic suspension gradually gains speed to very high RPMs by means of a special low-power electric drive, which sustains its turning. As a rule, a super-flywheel is located in a vacuum chamber to exclude air resistance. The super-flywheel is connected to a shaft of a generator by means of a powder clutch with a controlled slide to align the high RPMs of a flywheel with the slower RPMs of a generator. As the energy of a flywheel and its RPMs get lower, sliding in a clutch is also reduced. Such devices are manufactured as cabinet-type structures (Figure 4.24).

There are also UPSs of a similar type that are manufactured as separate modules; they can be assembled into larger groups in a standard container (Figure 4.25).

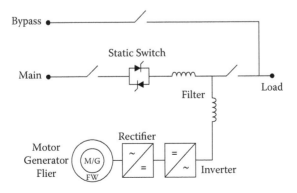

FIGURE 4.23
Diagram of a hybrid UPS based on a super-flywheel with a generator and inverter.

FIGURE 4.24
Hybrid-type UPS based on a super-flywheel and inverter for capacity starting from 250 kVA (single cabinet) to 1200 kVA.

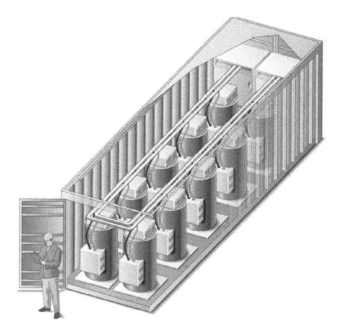

FIGURE 4.25
Hybrid-type UPS based on a super-flywheel and inverter, manufactured as separate modules with a capacity of 250 kVA, assembled into units with a capacity of 2.5 MVA in a standard container.

References

1. IEC 62040-3:2004. Uninterruptible power systems (UPS)—Part 3: Method of specifying the performance and test requirements.
2. Kulkarni, A. 2006. A hidden reliability threat in UPS static bypass switches. White paper no.96, American Power Conversion.
3. State Standard 13109-97. Electric energy. Compatibility of technical means, electromagnetic. Quality standards of energy in power supply systems of general purpose.
4. G5/4-1:2005. Managing harmonics. A guide to ENA engineering recommendation.
5. BS EN 50160:2010. Voltage characteristics of electricity supplied by public electricity networks.
6. IEEE Standard 519-1992. IEEE recommended practices and requirements for harmonic control in electric power systems. IEEE Industry Applications Society/Power Engineering Society.
7. Hoevenaars T., K. LeDoux, and M. Colosino. 2003. Interpreting IEEE Std. 519 and meeting its harmonic limits in VFD applications. IEEE Industry Applications Society 50th Annual Petroleum and Chemical Industry Conference, 2003. Record of conference papers, Sept. 15–17.
8. State Standard P 51317.3.2-2006 (MEK 61000-3-2:2005). Compatibility of technical means, electromagnetic. Emission of harmonic current parts by technical mans with consumed current not exceeding 16 A (in one phase). Norms and methods of tests.
9. State Standard P 51317.3.5-2006 (MEK 61000-3-5:1994). Compatibility of technical means, electromagnetic. Limitation of variation of voltage and flicker created by technical means with consumed current exceeding 16 A connected to low-voltage power supply systems. Norms and methods of tests.
10. State Standard P 51317.3.12-2006 (MEK 61000-3-12:2004). Compatibility of technical means, electromagnetic. Limitation of harmonic parts of current by technical means with consumed current exceeding 16 A, but not exceeding 75 A (in one phase), connected to low-voltage power supply systems of general purpose. Norms and methods of tests.
11. Standard UL1561. 1994. Dry-type general purpose and power transformers.
12. Standard UL1562. 1994. Transformers, distribution, dry-type, over 600 volts.
13. IEEE C57.110/D7-1998. Recommended practice for establishing transformer capability when supplying nonsinusoidal load currents.
14. Deshpande, K., R. Holmukhe, and Y. Angal. 2011. K-factor transformers and nonlinear loads. National Conference 2011, Bharati Vidyapeeth Deemed University, College of Engineering.
15. Handbook of American Power Conversion. 1994. APC. *The power protection handbook.* 1994. APC.
16. Dshochov, B. D. 1996. Features of electrical power supply of means of computer networks. *Industrial Power Engineering* 2:17–24.

5

Lead–Acid Accumulator Batteries

5.1 Background

Stationary lead–acid accumulator batteries (ABs) are the most important component of auxiliary DC power supply systems at substations and power plants. In Russia, the availability of ABs at substations is compulsory for voltage rates of 110 kV and higher, whereas in many Western countries an AB is the most important part of substations of all classes. Unfortunately, multiple regulatory documents—operational manuals and standards—cannot answer a number of questions that arise when working with ABs. This chapter is not meant to be a substitute for actual manuals and standards, but rather to supplement them with data necessary for understanding the arrangement of ABs and processes that take place inside them. This knowledge is compulsory for the correct selection of an AB type and its proper implementation.

The history of the lead–acid accumulator in its modern appearance starts in 1859 when French scientist Gaston Plante (an employee of the famous laboratory of Alexandre-Edmond Becquerel; Figure 5.1) took two thin lead (Pb) plates, put a piece of ordinary cloth between them (this part of an AB is now called a separator), and attached this "sandwich" to a wooden cylinder, which was then placed into a 10% solution of sulfuric acid (Figure 5.2).

The first accumulators of Plante had such a small capacity that there was no way to use them in practice. However, it was found that the bigger the surface of plates immersed into electrolyte was, the higher was the current rate. Then, lead plates with pores and ribs similar to fish gills appeared. It was found that if initially charged accumulator was discharged and then a current passed through it in the reverse direction and this procedure was performed several times, the capacity of the accumulator would increase significantly. This becomes clear when taking into consideration that it is not only pure lead (negative plate) that takes part in the reaction, but also its oxide and dioxide—PbO_2 (positive plate). Lead dioxide is formed on a positive lead electrode during accumulator charging. Multiple cycles of charging/discharging were required to achieve adequate results. This cycle was called plates forming and it took Plante more than 3 months.

FIGURE 5.1
Gaston Plante (1834–1889).

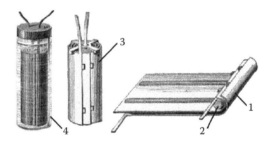

FIGURE 5.2
The first lead–acid accumulator assembled by G. Plante: 1—a "sandwich" made of two thin lead plates with a cloth stuffing between them; 2—wooden cylinder; 3—finished assembly of electrodes; 4—glass jar with sulfuric acid solution.

In 1878 French engineer-chemist Emile Alphonse Faure invented a new method of plate forming. He started covering ("spreading") plates with a red lead (Pb_3O_4) in advance, before the battery was assembled. During charging, the red lead was converted into peroxide (lead dioxide PbO_2) on the positive plate, while on the negative plate it was deoxidized, forming spongy lead. Moreover, due to multiple pores, the area of its surface increased significantly. The formation process conducted by Faure was much quicker and more efficient. As a result, Faure's batteries accumulated more electric energy than Plante's batteries with the same weight.

Faure's idea was developed further in 1881 by E. Volkmar, who created electrodes in the form of a filler grid, the cells of which held the red-lead dough extremely well. In the same year, another scientist, Sellon, received a patent for the production of grids made of a lead–antimony alloy.

Modern accumulators are manufactured using advanced technologies and materials, but the general principles of Plante, Faure, and Volkmar developed in the nineteenth century (Figure 5.3) remain unchanged.

FIGURE 5.3
One of the first plate-based accumulators—a prototype of modern devices.

5.2 Principle of Lead–Acid Accumulator Operation

The active material in a positively charged electrode (anode) of an AB is lead peroxide (dioxide), PbO_2, which is dark-brown in color, whereas the active material in the negatively charged electrode (cathode) is the pure (spongy) lead (Pb), which is light-gray in color. The electrolyte is a 25%–34% aqueous solution of sulfuric acid. The accumulator's capacity depends on the active surface of electrodes (plates) and on the number of the plates connected in parallel. There is always one more negative plate than the positive plates, since each positive plate is located between two negative plates. This is done in order to ensure uniform participation of both surfaces of a positive plate in the electrochemical reactions; if only one surface is working, a thin positive plate can warp and touch a negative plate.

The principle of operation of a lead–acid accumulator is based on the electrochemical reactions of Pb (cathode) and PbO_2 (anode) in a sulfuric acid medium. There are more than 60 reactions in a lead accumulator, but the overall chemical reaction (charging/discharging) is described by the following formula in accordance with a generally accepted theory of double sulfation:

$$PbO_2 + Pb + 2H_2SO_4 \rightarrow 2PbSO_4 + 2H_2O \text{ (discharging)} \quad (5.1a)$$

$$2PbSO_4 + 2H_2O \rightarrow PbO_2 + Pb + 2H_2SO_4 \text{ (charging)} \quad (5.1b)$$

During discharging,

$$Pb + HSO_4^- \rightarrow PbSO_4 + H^+ + 2e^- \text{ (on the cathode)} \qquad (5.2a)$$

$$PbO_2 + HSO_4^- + 3H^+ + 2e^- \rightarrow PbSO_4 + 2H_2O \text{ (on the anode)} \qquad (5.2b)$$

During charging,

$$PbSO_4 + H^+ + 2e^- \rightarrow Pb + HSO_4^- \text{ (on the cathode)} \qquad (5.3a)$$

$$PbSO_4 + 2H_2O \rightarrow PbO_2 + HSO_4^- + 3H^+ + 2e^- \text{ (on the anode)} \qquad (5.3b)$$

A partially soluble compound, $PbSO_4$, which is formed during discharging on both electrodes is called lead sulfate; this is why the theory describing this process is called the *theory of double sulfation*. Since, during discharging, sulfuric acid is used up for sulfate formation, its concentration is decreasing. In other words, the density of the electrolyte (so-called "specific gravity") is decreasing from 1.20–1.30 g/cm^3 (for different types) for charged accumulator to 1.01–1.03 g/cm^3. This, in turn, results in the reduction of voltage on the electrodes of such an accumulator from their initial rate of 2.10–2.22 V (depending on the initial concentration of sulfuric acid) to 1.95–1.70 V at the end of discharging.

During charging, a reverse process is taking place: Sulfuric acid is released into electrolyte solution from sulfates on the electrodes involving water (see Equation 5.1a, b); the density of the electrolyte and the accumulator voltage are increasing. During charging (almost at the end)—especially by higher current rates or at critical concentrations of lead sulfate on electrodes—electrolysis (decomposition) of water into hydrogen (near the cathode) and oxygen (near the anode) is observed. These gases form an explosive mixture in the air (about 4% of hydrogen by volume); this is why premises with ABs are potentially explosive. Moreover, water decomposition results in the reduction of its volume in the accumulator, so it has to be replenished from time to time.

It should be noted that there is a solution to this problem—that is, recombination plugs made by several manufacturers of accumulator batteries, in particular, by the renowned German company Hoppecke. The plugs of this company, called AguaGen® (Figure 5.4), contain a special catalyst that converts 98% of the trapped hydrogen and oxygen into water vapor that condenses on the plug's walls as small drops flowing back to the accumulator. Open acid ABs with liquid electrolyte and equipped with Hoppecke plugs are classified as maintenance free, since they do not require replenishment of water during their lifetime (25 years for GroE AB). As for the ventilation of rooms with ABs equipped with such plugs, they should be treated (based on the company's recommendation) as rooms with valve-regulated batteries (i.e., without forced ventilation).

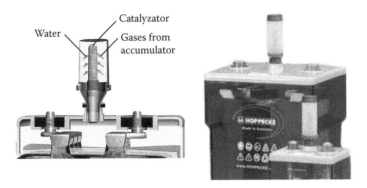

FIGURE 5.4
Accumulator of Hoppecke with recombination plugs (AguaGen⁰).

One other solution is automatic battery filling systems (BFSs). The systems include float-operated watering plugs (Figure 5.5), which replace the usual caps on each cell; hoses in transparent PVC pipes; connectors; and water carts. When the electrolyte level drops, the float-operated valves will open automatically. Every time the system is connected to a water cart, the electrolyte level will rise to the selected preset level. Such systems have been designed and manufactured by many companies: Vandapower, Battery Watering Technologies, Hawker, Emrol, Frotek, and others.

Due to the charging and discharging events in the course of AB use and because of volumetric changes, strength degradation of the anode's active mass (lead dioxide—PbO_2) and loss of mechanical and electrical connections between particles takes place. This leads to dissolving and guttering of the active mass. This phenomenon is enhanced by a considerable amount of oxygen bubbles emitting on the anode's surface during charging. This results in the deterioration of the accumulator quality, and finally it breaks down. In addition, there is a flaking of high-resistance lead dioxide parts from the

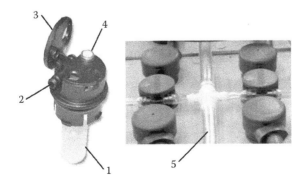

FIGURE 5.5
Float-operated watering plug of BFS: 1—float; 2—input tube; 3—cap; 4—valve position (electrolyte level) indicator; 5—hoses.

anode and electrophoretic translocation to cathode plates' exposed lateral edges, which can result in short circuiting between plates.

5.3 Effect of Electrode Sulfation on Accumulator Performance

Longevity of the active mass of a positively charged electrode is largely determined by conditions of lead sulfate ($PbSO_4$) crystallization during discharging. If lead sulfate residues are formed friable on the anode, this reduces damage of the active mass, since during charging this friable layer is converted into a solid active mass, which consists mainly of large crystals of lead dioxide. If the surface of the anode is covered with a hard layer of lead sulfate, then the crystals of lead dioxide (PbO_2) that are formed during this process are growing mainly as dendrites, which can flake off at the end of charging/beginning of discharging. The structure of the lead sulfate is largely dependent on the rate of discharging current (the higher the current rate is, the thicker the layer is) and temperature (the higher the temperature is, the lower is the density of lead dioxide layer).

In order to increase the strength of the active mass, electrodes are manufactured from lead alloys (vs. pure lead) containing 1%–2% antimony and other admixtures and often salts of calcium. However, the use of calcium salts leads to deterioration of accumulator qualities, such as significant and irreversible reduction of capacity as a result of deep discharging. This happens due to an irreversible reaction of calcium sulfate formation on the positive plate, which blocks the surface of the plate over time.

When discussing the issue of electrode sulfation, it should be emphasized that formation of sulfates $PbSO_4$ on accumulator electrodes is a natural and absolutely necessary process during discharging. Increased and excessive sulfation is harmful, since it results in the formation of a solid layer (vs. friable small-crystal sulfate), which consists of large crystals insulating the electrodes and preventing their contact with electrolytes. An accumulator with a high sulfate content has a dramatically reduced capacity during discharging. Another characteristic of increased sulfation is excessive gas emission and increased voltage of the accumulator in the very beginning of the charging process.

The following are the factors that result in deep sulfation of accumulator plates:

1. Too deep of a discharge. Theoretically, an accumulator can be discharged until full conversion of active masses of electrodes into a sulfuric lead and exhaustion of the electrolyte. However, discharging is stopped much earlier in practice, since sulfuric lead, which is formed during discharging (white salt, poorly soluble in the

electrolyte), has a low electric conductivity, which results in the sharp fall of accumulator voltage at the end of the discharging process. This is why discharging is stopped earlier, before 35% of the active mass is converted into a sulfuric lead. In this case, sulfuric lead will be evenly distributed in the form of tiny crystals in the residues of the active mass, which preserves sufficient electric conductivity to ensure 1.7–1.8 V between electrodes. In the case of a deeper discharge, partially irreversible changes take place on electrodes.

2. Systematic undercharging. This is when recovery of the active material—lead dioxide (PbO_2)—from lead sulfate ($PbSO_4$) (see Equation 5.1a, b) is not complete. The amount of active material is reduced and the amount of sulfate is increased, interfering with the balance of the chemical reactions.

3. Long-term storage without charging. When the AB is stored for a long time, it discharges itself. Similarly to ordinary discharging, the process of self-discharging results in consumption of the active material and formation of sulfate according to a reaction described by Equation 5.1(a, b). If the accumulator is stored charged for a long time, it needs to be recharged occasionally. If it is stored at –20°C, it should be recharged once a year for 48 hours at constant voltage rate of 2.27–2.44 V. When it is stored at room temperature, it should be recharged once every 8 months for 6–12 hours at constant 2.4 V. Storage at temperatures above +30°C is not recommended.

If the accumulator is too polluted by sulfate, it cannot be fixed and should be changed. Long-term overcharging at an increased current rate (2.4–2.7 V) is also dangerous for an accumulator. During this charging, water is intensively decomposing into oxygen and hydrogen. This process is accompanied by an excessive release of bubbles (water "boiling"), reduction of water volume, and reduction of electrolyte level. Though large amounts of moving bubbles can result in active material flaking from the anode plate, sometimes it is recommended to use this charging regime for a limited time period in order to perform mechanical cleaning of plates from lead sulfate crystals.

5.4 Classification of Lead–Acid Accumulators

Lead–acid accumulator batteries are classified based on several characteristics:

1. Structure of positive plates:
   - Large surface positive plates, which are also called "surface" or "Plante" (the name accepted in the West)—GroE

- Tubular plate (OPzS, PzS, OPzV, PzV, or PzB)
- Flat pasted plate (OGi, OGiV)
- Latticed—also called "Faure pasted-plate" or pasted plates (OP, VRLA)

2. Electrolyte condition:
   - With liquid electrolyte (GroE, OPzS, PzS, OGi, or OP)
   - With gel-like electrolyte electrolyte (OPzV, PzV, PzB, or OGiV)
   - With absorbed glass mat (VRLA)

3. Maintenance:
   - Flooded, vented lead acid (VLA) needing little maintenance, which requires water replenishment within operational life (GroE, OPzS, PzS, OGi, or OP)
   - Sealed, which is sometimes called "maintenance free"—does not require water replenishment within operational life (OPzV, PzV, PzB, OGiV, VRLA, or SVR)

4. Purpose:
   - Starter—to start internal combustion engines (SLI, ICE) and as a vehicle's power source
   - Tractive—for operation of loading machinery, self-propelled carriages and carts
   - For portable devices and equipment
   - Commercial stationary—for communication, power, and industrial facilities

Letters indicating accumulator class that are shown in parentheses were suggested in German standards Deutsche industrial norm (DIN) and are accepted worldwide today (Table 5.1).

5.5 Types of Lead–Acid Accumulator Battery Plates

Surface plates (Plante's electrodes) are generally similar to those electrodes that were used in the first models of accumulators. They feature a sheet of pure lead 10–12 mm thick with a large number of slots. These slots allow 8- to 10-fold increase of the plate's surface without increasing its physical sizes. Modern technologies of lead processing enable obtaining even more sophisticated surfaces (Figure 5.6). Due to electrochemical processes, a relatively thin active layer of lead dioxide is formed on the lead's surface. When the

TABLE 5.1

Description of Accumulator Battery Marking according to DIN VDE 0510-2: Specification for Electric Storage Batteries and Battery Plants

Marking	Description	Standard
GroE	Stationary batteries with surface positive plates in a traditional arrangement, flooded	DIN 40732
	Stationary batteries with surface positive plates in a narrow arrangement	DIN 40738
OPzS	Stationary batteries with tubular positive plates and a separator	DIN 40736 DIN 40737
OPzV	Stationary batteries with tubular positive plates and a separator, sealed with gel-like electrolyte	DIN 40742
OGi	Stationary batteries with flat pasted positive plates, flooded	DIN 40734 DIN 40739
OGiV	Stationary batteries with flat pasted positive plates., airtight, with gel-like or electrolyte	DIN 40737
GiV	Multiblock batteries with flat pasted positive plates	DIN 43534
SPzV	Special batteries with tubular positive plates, sealed, with gel-like electrolyte	DIN 43534 DIN 43539
PzS	Tractive batteries with tubular positive plates	DIN 43531-5
PzV	Tractive batteries with tubular positive plates, sealed, with gel-like electrolyte	DIN 43531-5

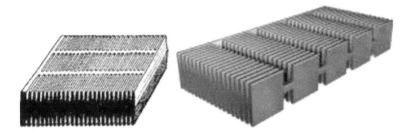

FIGURE 5.6
Modern arrangements of positive surface plate (Plante) with a big surface area.

accumulator is working, some lead dioxide falls down; however, new layers of lead dioxide are formed during charging.

This ensures longevity of operational life of surface plates, which can last as long as 15–25 years. Remember that surface plates are only used as positive electrodes in stationary accumulators, where reliability and longevity are much more important than specific energy. Negative electrodes in this type of accumulator consist of pasted plates, which are also called latticed or Faure plates.

If modern surface plates are almost similar to the first Plante's plates, latticed plates are almost similar to those of Faure–Volkmar.

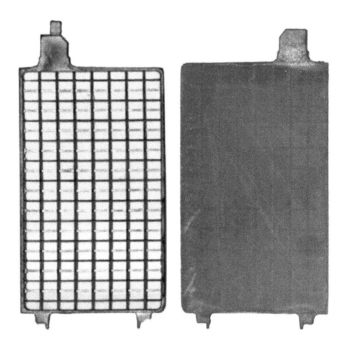

FIGURE 5.7
A grid made of lead–antimonous alloy (left) and a finished plate full with mass (right).

Latticed (pasted or *Faure)* positive and negative plates consist of profiled grids covered with a paste, which creates the active mass when the plate is being formed (Figure 5.7). The grids are usually cast of lead–antimony alloy, containing 5%–6% antimony and 0.2% arsenic, whereas the paste is made of lead powder and sulfuric acid with different bonding agents.

Accumulators with latticed plates have a high specific capacity (i.e., they need less space than accumulators with surface plates provided the capacity is similar), but they have a lower operational life compared to Plante's electrodes.

Pasted plates are widely used; they are implemented in both stationary and starter accumulators as well as in many other types of accumulators.

Box plates differ from latticed ones in that that they have additional box-shaped outer walls made of thin punctured lead sheets, preventing active mass from flaking off. They are about 8 mm in thickness. Box plates have the same specific capacity as grid plates; however, their mechanical strength is higher. Box electrodes are used as negative electrodes in combination with surface or tubular positive electrodes.

Tubular plates (Figure 5.8) consist of a rack cast from a lead alloy. Punctured plastic tubes (shells) or common compound casing is placed over the rods of this rack. The tubes are filled with active mass. The tubes are made of ebonite, vinyl plastic, synthetic fibers, and other materials; fiberglass lining is often used.

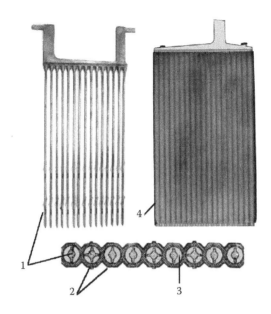

FIGURE 5.8
Tubular plates: 1—a rack made of lead rods; 2—punched plastic tubes (shells); 3—active mass; 4—finished plate.

Because the active material is held very well in the tubular plates, they are resistant to vibration during transportation and their specific capacity is 1.7–2 times higher than that of surface plates. They have a long life too— more than 1,000 charging/discharging cycles. These plates are used as positive electrodes in traction and sometimes stationary accumulators.

Regardless of the structure of electrodes, they are gathered in blocks of negative electrodes connected in parallel and blocks of positive electrodes connected in parallel (Figure 5.9), which are either put on legs at the bottom of a rectangular container (called a battery jar) or hung on internal lugs of the container.

The block of positive electrodes is inserted into the block of negative electrodes in such a way that the outside plates are always negative (there is one more negative plate than positive plates). There is a separator between each positive and negative plate, which insulates the plates, but allows the electrolyte through. There is an empty space at the bottom of the container, which is necessary to accumulate the guttering active material.

Starter and some other types of traction AB are assembled in monoblock jars (Figure 5.10). A monoblock is a single housing of a battery divided by partitions into three or six cells (depending on the number of accumulators for 6 and 12 V batteries, respectively), which are connected in series by means of external or internal jumpers.

Different types of accumulators use different combinations of electrodes, as shown in Table 5.2.

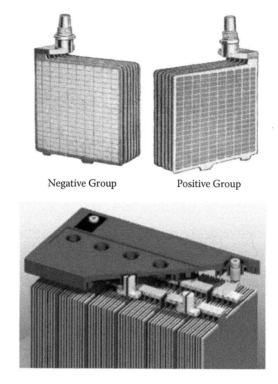

FIGURE 5.9
Units of positive and negative electrodes of a stationary battery.

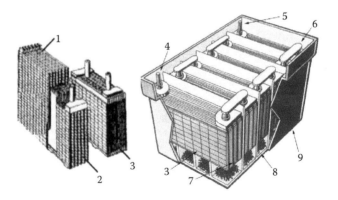

FIGURE 5.10
Monoblock design of 12 V traction and starter batteries: 1—block of negative electrodes; 2—block of positive electrodes; 3—block of electrodes of one cell, assembled; 4—positive pole terminal of the monoblock; 5—negative pole terminal of the monoblock; 6—jumpers connecting blocks of electrodes (cells) in series; 7—separators between plates; 8—fallen down active material of electrodes; 9—plastic housing of the monoblock.

TABLE 5.2

Combinations of Electrodes in Some Types of Accumulator

Accumulator Type	GroE (DIN40738)	OGiDIN40734 (DIN40739)	OPzSDIN40736 (DIN40737)
Capacity (A/hour)	15–600	12–3,500	40–12,000
and voltage (V)	2.23	2.23	2.23
Positive electrode	Plante	Pasted	Tubular
Negative electrode	Pasted	Pasted	Pasted

5.6 Types of Electrolytes

In addition to the accumulator with liquid electrolyte (flooded) discussed earlier, there are accumulator batteries with gel-like and absorbed electrolyte. The development of such electrolytes is connected with an attempt to create sealed accumulator batteries, which do not require maintenance.

In order to create an immobile electrolyte, it is thickened by a silica gel (SiO_2), which has high plasticity and fills both electrodes and a separator. Due to its viscosity, the thickened electrolyte is held in pores very well; it covers the plates and prevents flaking off of the active mass. When drying, the gel structure is covered by microcracks, which prevent the electrolyte evaporation from volatilization. The molecules of oxygen and hydrogen formed in the course of chemical reactions are held inside the gel, react with each other, and convert into water, which is absorbed by the gel. Thus, evaporation returns to the accumulator almost completely (95%–99%).

With a chemical imbalance as result of high storage temperature, high charging current rates, or the end of life cycle, all the gas molecules cannot be recombined, so some surplus gas leads to swelling of the cell walls as a result of increasing gas pressure up to a level insufficient to release through safety valves. The deformation varies from cell to cell and is greater at the ends where the walls are unsupported by other cells. Such overpressurized batteries may have normal electrical parameters under test, such as internal impedance and capacity, but they should be carefully isolated and discarded. One reason for this is lower performance of such accumulators, and another reason is a risk of explosion with spraying acid (as safety valves malfunction).

A separator in gel-like electrolytes is also unordinary. It is represented by Dura plastic with micropores, which is very tolerant of aggressive media due to aluminum additives. It also has low internal resistance, high thermal stability, and mechanical durability (vibration strength). This element is manufactured by only two specialized German companies. (All world-renowned manufacturers of accumulators are buying separators from them.) Gel-based accumulators sustain large amounts of charging/discharging cycles, can be stored discharged for a long time, have low self-discharging potential, and can be used in almost any position, which is very important for mobile installations and living spaces.

Accumulator batteries with gel-like electrolyte are marked as OPzV, PzV, PzB, or OGiV, depending on the type of plate implemented (see Table 5.1).

The drawbacks of an accumulator with gel electrolyte include high sensitivity to increased discharging voltage, at which this accumulator is quickly and irreversibly broken due to gel decomposition, and a sharp reduction of capacity due to low temperature because of excessive gel thickening.

In another method of immobilization, a separator made of fiberglass, which has high volumetric porosity and good wetting in sulfuric acid solution, is used. This separator not only separates electrodes, but due to a thin structure of fibers, it ensures keeping the electrolyte in pores as well as a high speed of oxygen transfer. The fiberglass separator, combined with thick assembly of the electrode block, also helps to reduce gutting of the active mass of the positive electrode and swelling of spongy lead on the negative electrode. This type of electrolyte is marked as AGM (absorbed in glass material) and is less sensitive to lower environmental temperatures than the gel.

Ordinary flooded accumulator batteries with a liquid electrolyte are called VLA (vented lead acid)—in other words, open—while accumulators that are gel-like or absorbed electrolyte are called VRLA (valve regulated lead acid) or SLA (sealed lead acid)—in other words, hermetic lead–acid accumulator batteries. Since *hermetic* is not be an absolutely correct word to call the accumulator, which has a safety valve to release surplus gas, it is called "air tight" in the Russian technical literature.

5.7 Dry-Charged Accumulator Batteries

Dry-charged accumulators differ from others in that their plates are charged (formed) before being assembled at the manufacturing plant; then they are washed thoroughly and dried by hot air. This accumulator comes already charged from the plant, but it has no electrolyte, so it cannot be used immediately after delivery. Dry-charged accumulators can be stored in a dry closed room at 5°C–30°C with tightly screwed sealed plugs for 1–2 years. To set them into operation, they are filled with an electrolyte, and the electrolyte's density is measured after 1–2 hours. If the density of the electrolyte is reduced by not more than 0.03 g/cm^3 compared with the density of the initial electrolyte, the accumulators can be used without additional charging. However, if the density of the electrolyte is reduced more than 0.03 g/cm^3, the accumulator needs additional charging before use.

5.8 The Capacity of an Accumulator Battery

This statement may seem obvious: Everybody knows what AB capacity means. But it is not that simple. The Standard IEC 60896-21 [1] of the International Electrotechnical Committee (IEC) and also IEC 60050 (international electrotechnical vocabulary—IEV) specify several different types of AB capacity:

- *Actual capacity:* quantity of electricity delivered by a cell or battery, determined experimentally with a discharge at a specified rate to a specified end voltage and at a specified temperature
- *Nominal capacity:* suitable approximate quantity of electricity used to identify the capacity of a cell or battery
- *Rated capacity:* quantity of electricity, declared by the manufacturer, that a cell or battery can deliver under specified conditions after a full charge
- *Shipping capacity:* quantity of electricity, declared by the manufacturer, that a cell or battery can deliver at the time of shipment, under specified conditions of charge

It is obvious that the different types of capacity mentioned here will have different rates (otherwise, why differentiate them?). Thus, when discussing the capacity of an AB, it is necessary to understand which capacity we mean. It is obvious that the rated capacity for determined specific conditions will not always match the nominal capacity, which is usually marked with the AB type and presented in promotional materials. At the same time, the actual capacity obtained as a result of discharging tests of ABs may be different from both nominal and rated capacity. It is clear that the most precise information about capacity can be obtained only during discharging tests. However, we have another surprise here. The same standard specifies different types of tests as well:

- *Accelerated test:* test in which the applied stress level is chosen to exceed that stated in the reference conditions in order to shorten the time duration required to observe the stress response of the item (battery) or to magnify the response in a given time duration
- *Acceptance test:* contractual test to prove to the customer that the device (battery) meets certain conditions of its specification
- *Commissioning test:* test applied on a device (battery) carried out on site to prove the correctness of installation and operation
- *Compliance test:* test used to show whether a characteristic or property of an item (battery) complies with the stated requirements

- *Endurance test:* test carried out over a time interval to investigate how properties of an item (battery) are affected by the application of stated stresses and by their time duration or repeated application
- *Laboratory test:* compliance test made under prescribed and controlled conditions, which may or may not simulate field conditions
- *Life test:* test to ascertain the probable life, under specified conditions, of a component or a device (battery)
- *Performance test:* test carried out to determine the characteristics of a machine (battery) and to show that the machine (battery) achieves its intended function
- *Type test:* conformity test made on one or more items representative of the production

In addition to these tests, American IEEE Standard 450-2002 [2] mentions one more type of test—*service test*—for flooded (VLA) stationary batteries. This test is conducted under "as is" conditions (i.e., without adjustments of discharging current and regardless of the age of the AB). Moreover, it is necessary for the test (as for many other types of tests meant for capacity measurement by discharging) to charge the accumulator battery fully. This so-called "full charging" is performed under the regime of equalizing charging at 2.3–2.4 V during 72 hours. However, IEEE Standard 1188-2005 [3] for stationary sealed (VRLA) batteries mentions that, during a service test, a full initial charging is not necessary if this is not specifically stated by the manufacturer of this type of AB as a normal occasional exploitation event.

In other words, the capacity of a stationary VRLA battery can be checked under "as is" conditions without adjustments of discharging current and without initial "charging." However, it should be noted that if the battery of accumulators is rejected based on minimum acceptable capacity criterion (80% of the rated capacity), this does not create grounds for filing a claim to a manufacturer. To prove the claim, a compliance test, laboratory test, or performance test should be conducted with initial charging of the AB under equalizing charging regime. This clarifies the importance of understanding features of different tests in relation to different types of capacity.

As mentioned previously, the criterion for accumulator rejection is a measurement obtained during actual capacity test, which is lower than 80% of the rated capacity. This rejection criterion is mentioned in international, American, and Russian [4] standards.

In accordance with references 2 and 3, a formula for capacity evaluation during discharging under direct discharging current regime for more than 1 hour is

$$C\ (\%) = [t_A/(t_S \times K_T)] \times 100\%$$

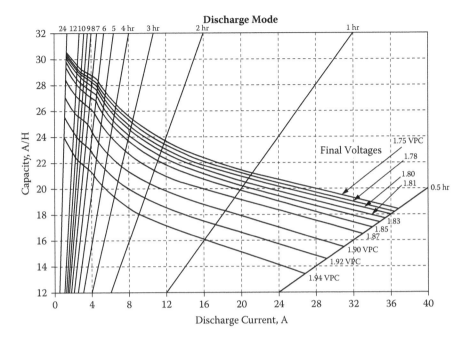

FIGURE 5.11
Discharging characteristics of V12V30T accumulator battery (schematic view).

where

C is accumulator (battery) capacity in percentage at 25°C

t_A is real time of battery discharging until minimum permissible voltage rating

t_S is estimated time of discharging (discharging regime), obtained from the manufacturer's chart (see Table 5.3)

K_T is adjustment coefficient for the electrolyte's temperature before discharging

Figure 5.11 shows discharging characteristics of Marathon sealed accumulator batteries (discharging regime) with time duration from 0.5 to 24 hours. Based on selected discharging regime and selected final discharging voltage rating, it is possible to obtain the rating of the discharging current, which should be maintained constant during discharging. The time within which the battery was able to provide this constant current until the moment of voltage drop to final discharging voltage is called real time of discharging.

Sometimes, manufacturers provide not only charts, but also schematic graphs of discharging characteristics, which are very convenient for practical use and allow one to obtain estimated measurements based on selected values.

5.9 Selection of an Accumulator Battery

In order to select the most suitable battery for a specific purpose, it is necessary to keep the following criteria in mind:

- Discharging regime and supplied capacity under this regime
- Occupied floor area
- Operation life
- Price

A standard estimated capacity of an AB is usually indicated for a 10-hour discharging regime. This capacity is marked as C_{10}. During short-term discharging regimes, the AB's capacity output coefficient will be less than one; this means that this capacity will be less than C_{10}, which is connected with a limited speed of chemical reactions on electrodes. The lowest capacity reduction at short-term discharging events (1- to 3-hour discharging regime) is provided by the batteries with surface positive electrodes (Plante); the next, in turn, are pasted. The batteries with tubular positive electrodes are the least suitable for short-term discharging regimes. This is why, when selecting them, it is necessary to use the batteries with a larger capacity rating to ensure necessary capacity reserve.

It should be kept in mind that, since the AB capacity is determined by capacity of its positive electrode (this is the reason that the name of the AB type is determined by the type of its positive plate), some manufacturers indicate in their specifications the capacity of the positive electrode versus AB capacity.

As for occupied floor area, sealed batteries (VRLA) are leading in this nomination. They are followed by (in order of area increase) flooded type batteries (VLA) with pasted electrodes and with tubular electrodes. The last place in this list belongs to batteries with surface type electrodes (Plante) (Figure 5.12).

The following factors influence the operation life of lead–acid accumulator batteries most of all: working temperature, deepness of discharging, and extent of recharging (length of charging at increased voltage). The operation life of any battery will notably reduce with the increase of any of these factors. However, if operating rules are properly observed, the most long-lived are the batteries that have surface positive electrodes (Plante) whose operational life is 20–25 years. The next in this list are the batteries with tubular electrodes (approximately 16–18 years). The operational life of batteries with pasted electrodes is 10–12 years. Most of the sealed accumulator batteries with gel-like or absorbed electrolyte have lower operational life, but they are cheaper. According to EUROBAT (Association of European Automotive and

FIGURE 5.12
A fragment of a substation battery for 230 V consisting of 106 lead–acid accumulators (with positive Plante's electrodes) connected in series.

Industrial Battery Manufacturers), sealed accumulators are divided into four classes based on their characteristics and operational life (Table 5.3 [5]).

VRLA batteries are the dominating batteries today for some applications, such as telecommunication backup, for example. The experience has been disappointing, however, because many published data report that capacity loss in VRLA batteries can occur at random intervals, which makes scheduling maintenance difficult. Some studies report that capacity begins to decline after as few as 2 years in service for a large enough portion of the series connected VRLA 6 and 12 V monoblocks to warrant concern. Such an effect has been named "premature capacity loss" (PCL). Practical experience indicates [6–10] that actual service life tends to be in the range of 4–8 years even on superior long life (>12 years of design life) types of VRLA batteries for various manufacturers and different discharge rates. It is true that VRLA batteries have a life considerably shorter than that of flooded VLA types.

TABLE 5.3

The Eurobat Classification

Lifetime in Years	Category	Abbreviation
≥12	Long life	LL
10–12	High performance	HP
6–9	General purpose	GP
3–5	Standard commercial	SC

Source: EUROBAT guidelines for the specification of valve regulated lead–acid stationary cells and batteries (May 2006).

Therefore, if we use these data to select, for example, a stationary battery for an auxiliary DC power supply system (230 V) at power plants and sub-stations, we can conclude that the batteries with positive surface electrodes (Plante) are the most suitable for this purpose (Figure 5.12), while hermetically sealed (VRLA) batteries are more suitable for portable devices with uninterruptable power supplies.

References

1. IEC 60896 Stationary lead–acid batteries. Part 21: Valve regulated types—Methods of test, 2004.
2. IEEE Standard 450-2002. IEEE recommended practice for maintenance, testing and replacement of vented lead-acid batteries for stationary applications.
3. IEEE Standard 1188-2005. IEEE recommended practice for maintenance, testing and replacement of valve-regulated lead-acid (VRLA) batteries for stationary applications.
4. State Standard 26881-86. Lead accumulator batteries, stationary. Main technical requirements.
5. EUROBAT guidelines for the specification of valve regulated lead-acid stationary cells and batteries (May 2006).
6. Moore, M. R., J. Vigil, S. McElroy, and G. Tsuchimoto. 2004. Real-time expected life and capacity on VRLA and flooded products unraveling and predicting code. IEEE 26th Annual International Telecommunications Energy Conference, INTELEC.
7. Hawkins, J. M., and R. G. Hand. 1997. Studies into the capacity retention behavior of VRLA batteries used in telecommunications applications. IEEE 19th International Telecommunications Energy Conference, INTELEC.
8. Karlsson, G. 1999. Premature capacity loss, an overlooked phenomenon in telecom batteries. IEEE 21st International Telecommunications Energy Conference, INTELEC.
9. Okada, Y., Y. Tsuboi, M. Shiomi, and S. Osumi. 2003. Premature capacity loss in VRLA batteries for telecom applications. IEEE 25th International Telecommunications Energy Conference, INTELEC.
10. Pavlov, D. 1993. Premature capacity loss (PCL) of the positive lead/acid battery plate. *Journal of Power Sources* 42 (3): 345–363.

6

Systems for Supervision of Substation Battery Continuity

6.1 Introduction

An auxiliary DC power supply substation system, shown in Figure 6.1, includes main and reserve auxiliary transformers, power battery, chargers, DC bus bars, and a distribution cabinet. This is an important substation system upon which the reliability of relay protection, automatic system control, and communication depend. According to reference 1, disturbances in this system may even lead to a full power system collapse.

A modern charger provides many different internal protective and signaling systems connected to emergency modes, while battery protection boils down, usually, to using a fuse. At the same time there is always the risk of failures in the contacts between the battery and the bus bar, in the links connecting the series of separate battery banks, in internal structure of the accumulators, and of the battery due to natural disasters, such as earthquakes. It is enough to take into account that a 230 V substation battery contains some 106 separate accumulators, connected together in series by means of more than 200 links; an interruption in any one of them can lead to complete battery malfunction.

6.2 Existing System for Supervision of Substation Battery Continuity

The Bender Company manufactures a device that can be used for monitoring the harmonics level in the DC network [2]. For a serviceable battery connected to a DC bus bar, the harmonics level is very low. It was assumed that the disconnection of the battery from the DC bus bar would dramatically increase the harmonics level produced by the charger's and pickups' output

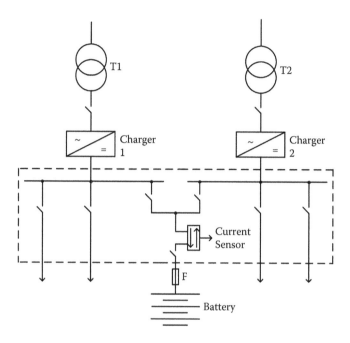

FIGURE 6.1
Typical single-line circuit diagram of a substation DC system.

relay in the Bender monitoring device. Unfortunately, the harmonics sum generated by the charger was high, depending on its output current (i.e., from the external load on the DC bus bar).

More to the point, in reality, modern chargers provided with large filter capacitors with a total capacitance in the range of 5,000–15,000 μF in output circuit determine a very low harmonics level on a DC bus bar, even with a disconnected battery.

We conclude from this that using harmonics level as the criterion for monitoring substation battery connectivity is not applicable.

Reference 2 describes a device for supervision substation battery connectivity based on a periodical pulsed increase voltage level on battery terminals and measurement current pulses passing through the battery.

Phansalkar et al. [3] describe a method for supervising the substation battery connectivity by injecting an audio frequency signal into the supervised circuit and measuring voltage drop on the circuit terminals on this frequency.

In Russo [4] and Wurst, Chester, and Morris [5], various devices for measuring the battery impedance as criterion for supervision of battery circuit connectivity are offered. However, Burkum and Gabriel [6] show that the conventional methods for measuring battery impedance are ineffective because of very low value of the impedance in power substation battery and

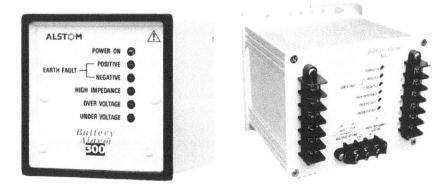

FIGURE 6.2
Device BA300 type (Areva) for constant monitoring of substation battery impedance, voltage, and ground insulation levels.

therefore very low value of AC voltage that needs to be measured. Measuring such low values of AC voltages in real substation conditions is problematic, as noted in Burkum and Gabriel [6]. Nevertheless, the Areva Company (previously Alstom) offered a special device: The Battery Alarm 300 [7] is specifically intended for measuring battery impedance (Figure 6.2). This device, in parallel with the battery, periodically connects a resistor by means of a semiconductor switch for a short period of time (50 μs). The resistor produces a short current pulse with a magnitude of 1 A. This current produces a small voltage drop across the battery terminals that is used for calculating the battery impedance.

Insofar as the battery charger contains large filtering capacitors in its output circuits, it is abundantly clear that the current pulse through the measurement resistor in the BA300 device will be formed not only by the battery, but also by the discharging of these filtering capacitors. In connection with this, Areva suggests inserting a chock, intended on full charger current, in the output circuit of each charger. Areva is somewhat reticent about where a consumer obtains such chocks on currents of 30 or 100 A and how much they will cost. Even so, their device itself costs about $1,000.

For the sake of justice, it is necessary to note that constantly monitoring the impedance of the battery can reveal not only the fact of a full break of the battery circuit, but also the deterioration of the general condition of this circuit even before its full break.

Also, the Battery Alarm 300 enables supervising additional parameters of the battery, such as voltage and the ground insulation level. However, for the author, the specific goal is limited only to the supervision substation battery connectivity. For this, the suggested solutions should be the simplest, most reliable, and most inexpensive so that it is possible to use a large number of them to cover all the batteries available in a power system.

6.3 Suggested Method for Supervision of Substation Battery Continuity

In distinction to the complexity (and consequently costliness) of the known methods for supervision of substation battery continuity, we are suggesting another method, based on the measurement current that permanently passes from the battery to the bus bar or from chargers to the battery (Figure 6.1) in a typical system. Even a fully charged battery continues to consume a small current (referred to as a "floating current") in a range of about 0.5–3 A from the charger, depending on the battery power and condition. Therefore, it is reasonable to consider that the current in a battery circuit reducing to a level less than 0.1 A, unequivocally testifies to breakage of this circuit.

It is necessary to pick a controller capable of giving the appropriate output signal at decreasing current in a supervised circuit lower than 100 mA. The problem with the selection of such a controller consists of the direction in which the current in a supervised circuit can change when reversing, as well as the change of value of a current in the circuit occurring in very wide limits: from 0.1 up to 100 A (i.e., by a factor of 1,000). Therefore, a high-sensitivity controller should be reliably protected from the influence of the high current value and should supervise a current in both directions.

According to these requirements, we constructed various units and tested some different systems for supervising the DC current in a battery circuit based on different principles.

6.4 Device for Supervision of Battery Circuit Based on Nonlinear Shunt

Using a nonlinear shunt allows considerable simplified supervision of current that changed in wide limits. The shunt in the device employs two back-to-back connected Schottky diodes. Forward voltage drop on one of the diodes (depending on the current direction) is changed according to the graph shown in Figure 6.3.

As can be seen from the graph, due to the nonlinear diode characteristic, the forward voltage drop is changed from 0.2 to 0.65 V (i.e., by a factor of 3 while the current is changed by a factor of 100). This feature of the diodes provides proper protection of sensitive input of the controller at high currents. On the other hand, enough high-voltage drops on diodes at small currents reduces the requirement of the sensitivity of the controller.

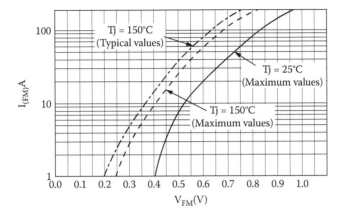

FIGURE 6.3
Dependence forward voltage drops on the Schottky diode STPS200170TV1 type from direct current (for different temperatures of semiconductor structures).

To test this idea, a circuit was put together modeled on the controller shown in Figure 6.4, using two standard devices produced by the Israeli Conlab company: DCT-3 and DCM-1. First, the insulated transducer with high-voltage insulation input circuit from output and also the converter input voltage ± 100 mV into standard output signal 4–20 mA was connected to them. Then, the controller itself with two programmed relay outputs was connected. The controller operated with input signals varying between 4 and 20 mA. For the two back-to-back connected Schottky diodes STPS200170TV1, manufacturer STMicroelectronics was used.

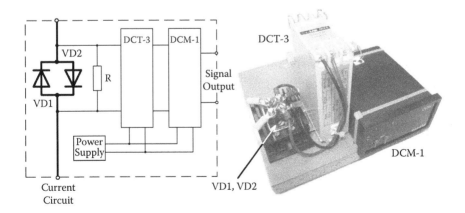

FIGURE 6.4
Model of the system based on nonlinear shunt for supervision battery circuit. DCT-3 and DCM-1: electronic transducers (Conlab, Israel); VD1–VD2: unit with back-to-back connected Schottky diodes STPS200170TV1 (STMicroelectronics).

Experimental examination of this system indicated that it was thoroughly passable. The output relay picks up at the reduction of the current in the supervised circuit below 50–60 mA and releases as current increases up to 130–140 mA. The presence of a high hysteresis in this case is a positive feature that increases stability of system operation.

At the same time an important deficiency was noted in connection with strong heating of the diodes unit at high currents. After having installed a small heat sink (Figure 6.4), the temperature the diodes unit achieved was about 70°C at 25 A and carried for a period of 15–20 minutes. From this it clear that, for operating at currents around 100 A, the diodes unit has to be combined with a large heat sink or a ventilator for forced heat sink blowing has to be used.

Another disadvantage of this system is the necessity for using a separate 24 V power supply for feeding the DCT-3 and DCM-1 devices. The cost of this system is about USD 600.

6.5 Using a Standard Shunt as Current Sensor

Using a standard linear shunt—100 A/60 mV or 100 A/100 mV, for example—and current pickup level 0.1 A, the sensitivity of the controller has to be 10 μV (unlike the previous example, where the sensitivity of the controller could be as low as hundreds of millivolts). It is unlikely that a controller available in the market has so high a sensitivity.

At our request, the Conlab company gave us another system for testing (shown in Figure 6.5) that met the requirements mentioned earlier for using a standard shunt. During the test, this system exhibited the following results: pickup current levels of 60 to 80 mA (through the shunt) and release current levels of 120 to 160 mA. The same hysteresis is a positive property of such a system because its presence increases the system stability. One of the disadvantages of this system is the necessity of using an external 24 V power supply for feeding the transducers. The cost of this system is about USD 600.

The MSCI-LCD-S type serial line controller, manufactured by Megatron, Israel, with some slight modifications to meet our requirements appeared to be a very successful variant and demonstrated very stable operation with high sensitivity (Figure 6.6). This controller practically did not react with the AC component when measuring the input and therefore had a high noise stability and could feed from the 230 V network.

It was noticed that the controller MSCI-LCD-S type lacked detection of the polarity of the input signal. In the modified variant of the controller, the measuring input is protected against the high voltage applied at high currents carried through the shunt and also protected against voltage polarity

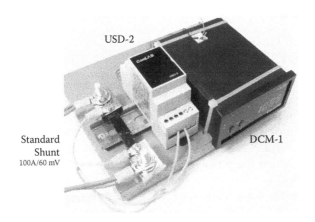

FIGURE 6.5
Model of the system based on standard shunt for supervision battery circuit: USD-2—electronic transducer (input: ±200 mV, output: 4–20 mA); DCM-1—programmed electronic transducer for 4–20 mA with relay output (Conlab, Israel).

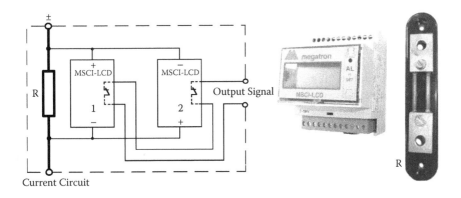

FIGURE 6.6
System for supervision battery circuit based on standard shunt and two universal controllers with relay output MSCI-LCD-S type (Megatron, Israel).

reversals applied to measuring the input during changes of the current direction in the shunt.

A single problem is the necessity of using two identical controllers connected in opposite polarity to the shunt for measuring the currents, proceeding in both directions, and connecting the normally closed contacts of its output relays in series. Thus, the output signal will appear only when the current is lower than 0.1 A in both directions. Considering the small cost of a single controller (about USD 130), it turns out that, when two such controllers are used, the total cost is much below that of the second system provided by Conlab.

6.6 Using the Hall-Effect Sensor in Systems for Supervision of Battery Circuits

The monitoring systems considered previously demand the insertion of additional elements (diodes, shunt) in a power circuit (cable) connecting the battery with the DC bus bar. However, there is a variation in which there is no need to cut the circuit (cable) or insert additional elements. This variation is based on using a Hall-effect sensor (transducer) in the form of a framework through which the power cable connecting the battery and bus bar is passed (Figure 6.7).

Some companies, for example, CR Magnetics [9], have proposed DC current relays with built-in Hall sensors. However, from the answers that we received from the company, it seems that such devices cannot provide carrying currents varying between 0.1 and 100 A and pickups at currents below 0.1 A.

We therefore undertook the development of the current relay of our own design based on separate Hall transducers, such as HAL50-S, which is manufactured by the Japanese branch of the LEM company and designed for a currents up to 150 A, and the AM22D type controller, manufactured by the Israeli company Amdar Electronics & Controls.

This type of Hall transducer consists of a built-in electronic amplifier and two potentiometers that deduce outside for adjusting the amplifier characteristics. Due to the amplifier, the output signal of the transducer is as high as ±4 V at nominal current of ±50 A. We assumed that such a high output signal at nominal current would provide enough high-level signal at current of 0.1 A also.

Unfortunately, even at zero current, the output signal level drifted and its instability exceeded the level of the functional output signal at a current

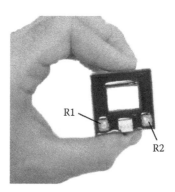

FIGURE 6.7

Hall-effect transducer HAL 50-S type with built-in electronic amplifier for measurement of DC current: R1, R2—potentiometers for amplifier adjusting.

of 0.1 A. Thus, we would not be successful in receiving the comprehensible signal applicable for use in our supervision system from such a device.

6.7 Newest Developments and Prospects of Their Application

More recently, in many cases, after the author had made inquiries to the manufacturers addressing the problems noted, some companies improved the devices and now offer new products that are better suited for the parameters for monitoring a DC system. For example, transducers type CR5211-2 (advertised by CT Magnetics) are ostensibly specially intended for monitoring DC system circuits without the external shunt and are ideally suited to our needs.

The transducers are advertised as being mounted directly on the cable. The sensing element of this device is not the Hall element, which is poorly suited to monitoring small currents, and a high-frequency transformer the magnetic condition of which varies depending on the magnitude of the direct current flowing through the monitored cable. The manufacturer's explanation is that "the ranges 2 to 10 Amp utilize an advanced Magnetic Modulator technology" [9]. This principle of such a sensing transducer reminds one of the magnetic amplifiers that were widely used in automatics in the 1950s and 1960s. List price of this device is USD 125. Another device of an analogous principle is offered by the Powertek UK company: the Current Sensor CTH type 2 for current range of 0–5 A, which costs about USD 360. It is only a transducer, with a transformed input signal in an output voltage. If we add to it a voltage relay—for example, SM125 024 V DC 4V (Carlo Gavazzi)—timer H3DE-S1 (Omron Industrial Automation) and a miniature Deutsch industrial norm (DIN) rail power supply 24 V DC—for example, type DSP10-24-PSU (TDK Lamda)—the cost of this complete set will increase to USD 300 for first example of the current sensor and USD 520 for the second one (Figure 6.8).

The previously mentioned Megatron (Israel) company, upon request of the author, has also modified its MSCI-LCD-S controller (depicted in Figure 6.6) and has made it insensitive to the polarity of current flowing through the shunt. This enables using only one such modified controller (MSCI-LCD-SM), instead of two, at a cost of about USD 130. Considering the increase in the device's cost, there was an idea of adding the additional controller for monitoring the charger condition of the DC monitoring system (Figure 6.9).

In all chargers, there is the built-in shunt intended for the measurement of an output current. The controller MSCI-LCD-SM can also be connected to this shunt. This controller watches the serviceability of the charger.

As has been shown in our experience, there is an assemblage of types of internal faults of chargers, at which no signals are generated. Thus, the

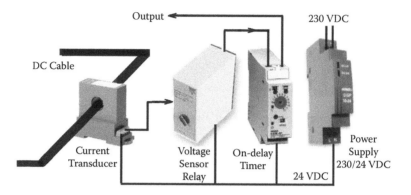

FIGURE 6.8
Structure of the DC circuit monitoring system based on novel type of small current transducer.

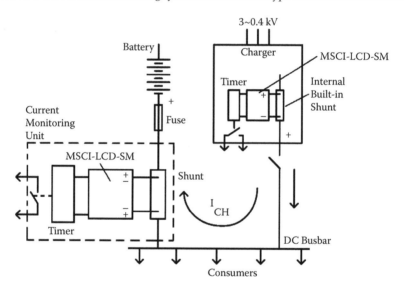

FIGURE 6.9
Circuit diagram for monitoring DC system.

battery without the charger feeds the loads until its voltage does not decrease to a certain level and the supervision relay (voltage relay) generates a signal about a fault. Use of the MSCI-LCD-SM controller would allow increasing the reliability of the charger monitoring and even completely canceling the voltage monitoring. The addition of a simple timer—for example, H3DE-S1 (cost of about USD 30)—with an ON-delay function and a time delay of 10–15 seconds to the circuit prevents output from spurious signals at switching in DC systems.

The main current monitoring unit and shunt are placed in different ways (Figure 6.10). The additional complete set of the controller with the timer can

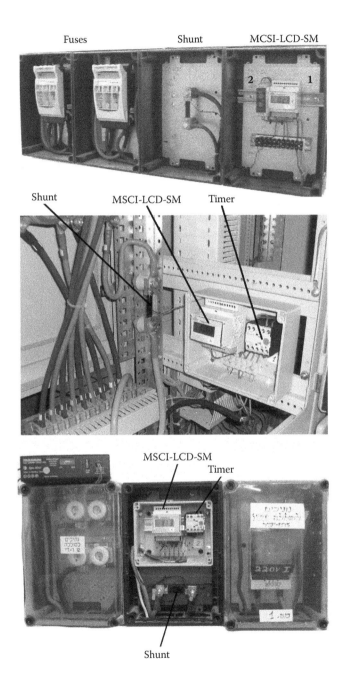

FIGURE 6.10
Main DC monitoring unit mounted differently in some substations.

FIGURE 6.11
Additional current monitoring unit mounted in charger: 1—controller; 2—timer.

be mounted directly in the charger (Figure 6.11). The sensitive input of the controller is connected directly to the regular internal shunt of the charger.

Additional experiments have shown that the system also works stably without the timer; therefore, it will be excluded from the scheme in the future.

6.8 Conclusion

On the basis of a comparative estimation of parameters and test results of some variations of systems for supervision circuits of the substation battery 230 V that were discussed earlier, we have come to the conclusion that, from the standpoint of the greatest stability, greatest reliability, and least cost, the system of choice is that based on a standard shunt and modified MSCI-LCD-SM type controller, manufactured by Megatron. This is the system that we recommend for wide use in substations and power stations.

References

1. Skok, S., S. Tesnjak, and M. Filipovic. 2003. Risk of power system blackout caused by auxiliary DC installation failure. Proceedings of the IASTED International Conference "PowerCON 2003—Special Theme: Blackout," New York, December 2003.
2. Ripple Detector RUG1002Z. Insulation and voltage monitoring of DC system RGG804. Operation manual. Bender GmbH.
3. Phansalkar, B. J., P. N. Tolakanahalli, N. R. Pradeep, and S. K. Saxena. 2005. Method and system for testing battery connectivity. US Pat. 6931332, August 16.
4. Russo, F. J. 1999. High sensitivity battery resistance monitor and method. US Pat. No. 5,969,625, G01R31/36, October19,1999.
5. Wurst, J. W., T. Chester, and C. Morris. 1994. On-line battery impedance measurement. US Pat. No. 5,281,920, G01R31/36, January 25, 1994.
6. Burkum, M. E., and C. M. Gabriel. 1987. Apparatus and method for measuring battery condition. US Pat. No. 4,697,134, G01R31/36, September 29,1987.
7. Battery Alarm 300. User manual BA300. Areva.
8. Direct Current Sensing Relay CR5395 Series. Data sheet. CR Magnetics, Inc.

7

Backup of Substation DC Auxiliary Power Systems

7.1 Characteristics of Backup Layout with Diodes

DC auxiliary power supply systems (DCAPSs) represent an extremely important system of substations and power plants, the reliability of which enables the ability of substations and power plants to perform their functions in the energy system. This is why strict requirements are applied to these systems and the problems and solutions in this field are discussed during international scientific and practical conferences [1].

One of the options to increase the reliability of the DC auxiliary system is its backup. A solution to this problem can be the use of two groups of accumulator batteries (ABs) and battery chargers (BCs). Each of these groups can consist of one AB and one or two BCs and be connected to its specific bus section on the DC control and distribution board (DC CDB). The sections are united by means of a switch, which remains open under normal modes of operation. These bus sections supply power to arrangements of DC feeders, which in turn supply power to relay protection and automatic (RPA) devices. According to reference 2, DCAPSs should include a manual input of backup by means of a contact breaker (or manually driven switches). However, in reference 3, automatic backup of DC lines is also introduced. According to 8.6 in reference 3, the automatic switching-on of backup power supply of RPA, DC feeders can be connected to DC CDB sections through splitting diodes. These diodes are rather rare in Russia and the Commonwealth of Independent States (CIS) countries, but are widely used in DC power supply networks of RPA in Western countries. One of the options of so-called "hot" backup of RPA DC power supply by means of diodes, which is widely used in some countries, is shown in Figure 7.1 (simplified).

In this layout, all RPA devices (main and backup protection) are split into several groups (not more than three or four per substation). Each of these groups receives power from the DC bus sections simultaneously—DC1 and DC2 through DC4 diodes, which provide splitting between positive and negative pole terminals of these two sections of DCAPS buses. This technical

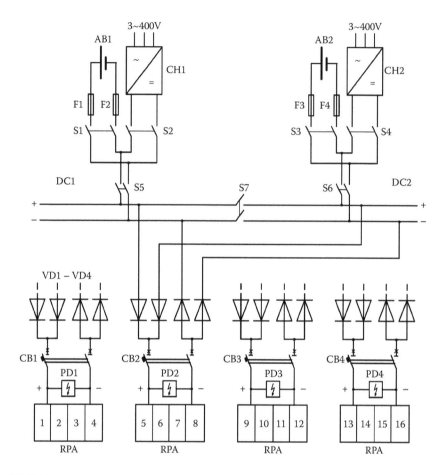

FIGURE 7.1
Simplified layout of power supply backup of RPA groups by means of diodes.

solution is simple and efficient, does not include commutation devices, and provides reliable backup of RPA power supply lines.

However, regardless of all its positive characteristics, this layout is far from ideal. Practical application of this layout finds multiple cases of RPA break-downs due to the connection of the voltage of DC1 and DC2 in series through diodes. As a rule, this happens during setup or maintenance in the DCAPS or in case of the accidental closing to the ground of the circuit connected to one pole terminal of DC1 and subsequent accidental closing to the ground of the circuit connected to the opposite pole terminal of DC2. In this case, the RPA receives 460 V, which destroys all the devices affected by this volt-age. Considering the high price of modern digital protective relays (DPRs) as well as the danger of unpredictable consequences of accidental large-scale failures of RPA devices, the danger for the energy system using this layout becomes obvious.

Analysis of large-scale failures of DPRs after an event like that described earlier showed that modules of power supply sources and modules of logic inputs are affected more often. In most cases, these breakages are limited to the burnout of varistors in input circuits, but there are cases of breakages of electronic components installed behind these varistors. Varistors are nonlinear resistors whose resistance depends on the voltage applied to them. They are designed to protect power supply lines and logic inputs of DPRs from short overload impulses (voltage spikes), which emerge in the DCAPS. Since the steepness of the volt-ampere characteristic (VAC) of varistors is far from ideal, they are selected in such a way as to be located at a working point as far from the area of high nonlinearity as possible under nominal working voltage and the current should not exceed 1 mA. These conditions will not allow the voltage, which is constantly distributed on a varistor, to exceed its maximum permissible level; the varistor will not heat up, thus maintaining its longevity. This is why varistors of the 431 class (e.g., type 10D431K) with a clamping voltage of 710 V are selected for operating a power supply of 240 V.

When a short pulse with a voltage amplitude exceeding 710 V is coming to the input of the circuit protected by a varistor, it transfers to a steep nonlinear area of VAC. Its resistance decreases sharply and it short-circuits the input, dispersing the energy of the pulse. When the pulse is over, the varistor returns to its initial condition, if the energy of the pulse did not exceed the permissible level of energy dissipation by varistor. When the varistor receives voltage exceeding its nominal rating, but lower than the rating of activation, the varistor will "get stuck" in the nonlinear area of the characteristic, which results in its extreme heating. Moreover, electronic parts located behind this resistor will also be affected by high voltage. This regime may result in full destruction of the varistor and its transfer to a mode of continuous short circuit within time frames from a fraction of a second to several seconds, depending on its power and its actual VAC. However, sufficient current continues flowing through, resulting in activation of an automatic circuit breaker, which disconnects the power supply of this group of already broken DPRs.

Actually, this regime can be prevented from operating a DCAPS by means of a continuously closed contact breaker S7, but it is unlikely to be accepted, since short circuit in circuits of one of the bus sections DC1 or DC2 can result in full loss of the auxiliary power supply. This is the reason two accumulator batteries working in parallel are not allowed [2].

7.2 DPR Protection Device

It is possible to protect the DPR from damage in such a layout of the auxiliary power supply by means of special protection devices connected between the

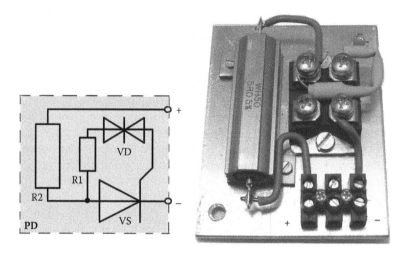

FIGURE 7.2
A diagram and arrangement of a sample protection device (PD).

"plus" and "minus" of input circuits of the DC feeder that feeds groups of DPRs (Figure 7.1). Each protection device has a very simple design and is assembled on an aluminum panel 80 × 65 mm (Figure 7.2).

This device includes thyristor VS, type MCO 100-12io1; suppressor VD, type 1.5KE300A; and current limiter on a resistor R2 (10 Ω resistance and 50 W power), as well as a low-power resistor R1 (200 Ω) in the circuit of the thyristor's gate. It would be good to add one more small resistor (300 Ω) connected between a cathode and gate of the thyristor, which will increase its immunity.

Thyristor VS is closed (not in a conducting mode) under normal working voltage, and current does not flow through it or through the current-limiting resistor R2. When a double supply voltage emerges, resistance of the VD suppressor dramatically drops (at 300+ V), and the gate of thyristor VS receives current that results in its opening (transferring to conducting mode). The thyristor opens and turns on a low-resistance resistor R1 between the "plus" and "minus" of input circuits of the DC feeder, leading to ~40 A current flow through this resistor (and also through a corresponding circuit breaker [CB]). This results in the immediate activation of a corresponding CB, cutting off a corresponding group of DPRs and protecting it from damage.

The main switching element (thyristor, type MCO 100-12io1) is designed for a maximum current of 156 A and pulse current of 1400 A (during 10 ms), and its maximum withstanding voltage reaches as high as 1200 V. The choice of a thyristor with such big reserves is stipulated by a necessity to ensure a high level of reliability of a device. The time of opening a thyristor does not exceed several microseconds, which ensures very quick (within several microseconds) response of the device; in other words, it ensures its high efficiency. The price of a kit of elements for such a device is about USD 50.

7.3 Automatic Reclosing Device for DCAPS

The technical solution described previously can be implemented in an existing DCAPS. In newly designed DCAPSs, it is recommended to avoid implementation of splitting diodes, which represent a potential danger, and rather to design it as two split bus sections equipped with automatic reclosing (AR) as shown in the example in Figure 7.3. The suggested AR system consists of two low-power auxiliary relays, K2 and K4, as well as two relays, K1 and K3, with powerful contacts designed for switching of direct current at 240 V.

The circuit is off and consumers are disconnected when contact breakers S10 and S11 are open, and circuit breakers CB1–CB4 are off. The order of

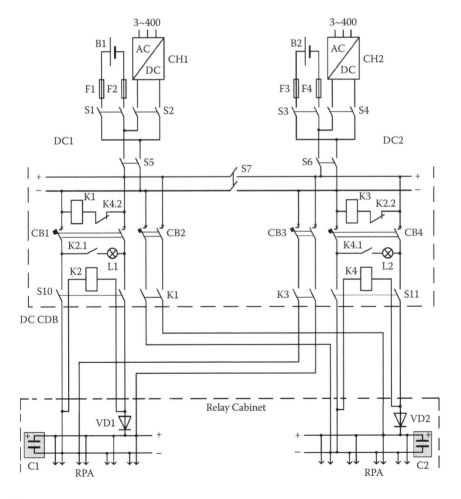

FIGURE 7.3
Suggested layout of automatic reclosing (AR) in a DCAPS.

supplying power to consumers is as follows: CB1–S10–CB4–S11–CB2–CB3. Thus, all consumers of the DC feeders will be powered, K1 and K3 contactors will be open (deenergized), and the circuit is in normal working mode. Warning lights L1 and L2 (or any other type of alarm) are on, which indicates that circuits of voltage control on K2 and K4 relays are in good shape.

In the event of a power failure on one of the sections (DC1 or DC2), disconnection of one of the circuit breakers (CB1 or CB4), or wire breakage in the supply line connecting DC CDB and DC feeders, one of the corresponding relays, K2 or K4 releases, and K1 energizes a corresponding power relay K1 or K3 and by its adequate NO contacts provides backup power supply from the uninjured bus section.

During the switching of AR relays, there can be a short-term (40–60 ms) voltage dip in the power supply of the DPR. This delay in the DPR power supply does not actually cause problems. According to the international standard [4], the DPR should tolerate dips in power supply from 10 ms to 1 s (at the manufacturer's discretion). The actual values of tolerable voltage dips that we measured for many DPR devices of foreign manufacturers amounted to 1.2–3.8 s. Even the Russian DPR, according to technical requirements, should tolerate power supply delays of up to 500 ms without malfunction reloading. This is more than enough for AR action.

However, DC circuits supply power to many other auxiliary electromagnetic relays, the deenergizing of which during AR action is not acceptable. In order to prevent voltage dips in the auxiliary power supply, the AR circuit includes a special unit C, which sustains power supply to RPA devices during this time (Figure 7.4). This unit is designed as part of a high-capacity capacitor (3700 μF, 400 V) equipped with a fuse with high interrupting duty, a resistor, and a diode. In this layout, the charging of the capacitor when it receives power is achieved through a resistor, which limits current to 1–2 A, and discharging is performed directly through a diode VD. Due to this connection layout, the capacitor does not create current steps, which can cause activation of circuit breakers, but at the same time it provides the rather high discharge current necessary to supply consumers during several tens of milliseconds spent switching relay of the AR device.

Any type of portable DC contactors can be used as K1 and K3 contactors in the device. These contactors are manufactured by many leading companies. For example, single-pole contactors of the ABB Company include three contacts connected in series and a built-in powerful permanent magnet to extinguish the DC arc (Figure 7.5).

More advanced technological solutions are implemented in DC contactors of the Omron Company (Figure 7.6). These contactors include a combination of all known methods of extinguishing of DC arcing:

- Serial connection of several contacts
- Installation of contacts in a hermetically sealed chamber filled with gas under excessive pressure

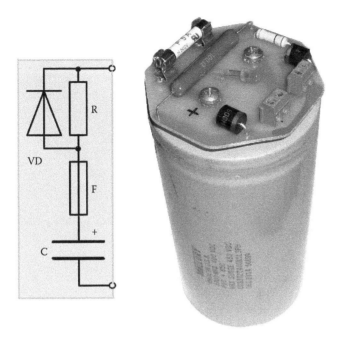

FIGURE 7.4
A unit that sustains power supply to RPA devices during AR switching

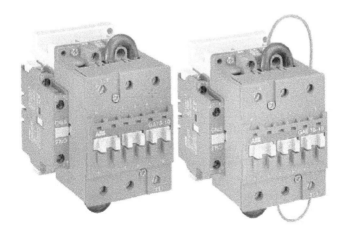

FIGURE 7.5
DC contactors, types GA75 and GAE75, of ABB Company.

FIGURE 7.6

Gas-filled hermetically sealed one-pole DC contactor G9E series (Omron) with arc extinguishing magnets and one NO double-break contact.

TABLE 7.1

Main Features of DC Contactors, Series G9E

Main Features	G9EA-1 (B)	G9EC-1 (B)	G9EB-1 (B)
Switching DC voltage, V	400	400	400
Maximum switching current, A	100	200	25
Continuous current through closed contacts, A	60	200	25
Short-term current (10 min) through a closed contact, A	100	300	50
Maximum cutoff current of short circuit, A	600 (300 V)	1000 (400 V)	100 (250 V)
Maximum cutoff current of overload, A	180 (400 V)	700 (400 V)	50 (250 V)
Electrical strength of insulation, kV AC, 1 min	2500	2500	2500
Weight, g	310	560	135

- Implementation of a permanent magnet to push the arc from the intercontacts gap

The main features of contactors of this series are listed in Table 7.1.

Diodes VD1 and VD2 prevent feeding of the short circuit (SC) location from the backup section of busses when AR is activated as well as discharging of the capacitor through the SC point. Any diodes for 100–250 A current with tolerable reverse voltage of 1200 V manufactured in housings convenient for installation in cabinets can be used in a device (Figure 7.7).

These are diodes of the following types: VSKE196/12, VSKE250/12, VS-UFL230FA60, HFA200FA120P (Vishay); MEO450-12DA (IXYS); T110HF120 (International Rectifier); and others. Diodes should be installed on the inner wall of a metal cabinet or, if this is not possible, they should be installed on an aluminum 5–6 mm thick plate with dimensions not less than 200 × 200 mm.

Recently, several leading manufacturers started manufacturing DPRs with two independent internal power supplies that work in parallel in the mode

FIGURE 7.7
Modern diodes for 100–250 A current and 1200 V, manufactured in housing and convenient for installation in a DCAPS cabinets.

of hot backup. These DPRs can receive power from two different bus sections of the DCAPS without any additional technical solution. However, there are only a few DPRs of this kind and they are still very expensive. Also, when there are three or four DPRs in one relay protection cabinet with double power supply, we have six to eight power supplies in one cabinet. And we can ask a logical question as to whether this solution makes any sense. It would be better to manufacture DPRs without built-in power supply; all the devices located in this cabinet will receive power from two highly reliable power supplies, which will serve the whole cabinet [5,6]. But this will come in the future. Now, we have to work with a rather diversified DPR device for which we need to find technical solutions today. Additionally, using the special DPR with a double built-in power supply does not solve the problem of power supply to auxiliary electromagnetic relays.

Technical solutions offered in this chapter can serve as a foundation for designing inexpensive but very reliable DCAPSs, which can be used to supply power to RPA devices of any type.

References

1. New solutions to build highly reliable DC systems for energetic installations. World trends. First International Scientific and Practical Conference, February 17–18, 2005, Moscow.
2. STO 56947007-29.120.40.093-2011. A guide to design DCAS PS ENES. Typical designs. OJSC "FSK ES." Introduced on June 1, 2011.
3. STO 56947007-29.120.40.041-2010. Direct current auxiliary systems of substations. Technical requirements. Organizational standard. OJSC "FSK ES." Introduced on March 29, 2010.
4. IEC 60255-1. Measuring relays and protection equipment—Part 11: Voltage dips, short interruptions, variations and ripple on auxiliary power supply port.

5. Gurevich, V. I. 2007. Microprocessor relay protection: The present and the future. *News in Electrical Energy Industry* 5:39–45.
6. Gurevich, V. I. 2010. New concept of construction of microprocessor based relay protection. *Components and Technologies* 6:12–15.

8

Insulation Problems in Substation DC Auxiliary Power Supply

8.1 Faulty Actuation of Relay Protection at Grounding of One of the Poles of Auxiliary DC Power Supply

To expand the range of operational voltage, logic inputs of digital protective relays (DPRs) are often designed to be able to engage at very low (in relation to nominal rating) voltage rates (e.g., 60–80 V at nominal rating of 230 V). In the newest installations, a wide range of acceptable operational voltage rates is divided into several subranges, which can be selected by a consumer depending on specific conditions of use.

Actuation voltage of all auxiliary electromagnetic relays is always lower than the nominal voltage. This is done to increase reliability of relay actuation and reduce time of actuation. These two features emphasize the danger of faulty actuation of both the DPR and the electromagnetic relay.

When one of the terminal poles is closed to the ground (either positive or negative), a current impulse with voltage rate equal to half of the battery's voltage (i.e., around 115 V) appears in the DC system, which is stipulated by capacitive discharge of wires of the DC system (Figure 8.1). As discussed in Gurevich [1], this discharge creates conditions for self-actuation of logic inputs of microprocessor protection devices if the actuation threshold of logic inputs is lower than half of DC circuit voltage (i.e., 115 V). Research of the reasons for several emergency shutdowns has shown that the capacitance and duration of discharge current pulse is sufficient to actuate not only logic inputs of microprocessor relays, but also some types of auxiliary electromagnetic relays with nominal voltage of 220 V (these relays were actuated at 70–90 V).

Considering that the discharge current pulse is limited in time, it is technically possible (by software tools) to introduce an additional programmable timer into the logic circuit of a microprocessor relay, which will delay the passage of input signal by 20–40 ms. This timer will act like a filter, preventing actuation of the microprocessor relay under short-term current impulses during capacitive discharge of the circuit. It is also possible to use an RC

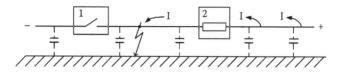

FIGURE 8.1

A diagram of capacitive discharge of a DC circuit through logic input of a protection relay: (2) with open external control contact; (1) when closing to the ground.

chain connected to relay inputs, which will perform a similar function. However, we need to determine if the increase of relay protection response time is acceptable at all. We think that this slowdown of relay protection response is not acceptable.

Taking everything mentioned previously into consideration, we suggest another solution to the problem, which is based on increasing the lower threshold of logic inputs actuation to the rated level that exceeds half of the circuit's voltage. A voltage of 150 V was selected as such a level. To realize this idea, we suggest using a very simple module (Figure 8.2), which consists of two electronic components: a Zener diode with nominal voltage of 150 V and nominal current of 5 mA and a powerful thyristor for 7.5 A current and 800 V with 5 mA gate current. At voltage rates lower than 150 V, a device rests closed (not conducted mode), so voltage pulses up to 115 V emerging in a DC circuit when closing to the ground do not pass to the logic inputs of the protection relay.

When control contact S is closed to a Zener diode, a voltage of 230 V will be applied in a step, and it will open immediately, letting current through it to the circuit of the thyristor gate. The thyristor will open (transfer in conducted mode) and shunt the Zener diode and gate circuit with its low direct resistance. Now, total current is passing through the thyristor's circuit

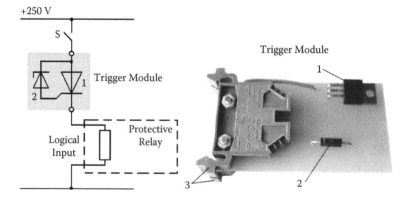

FIGURE 8.2

Connection diagram and arrangement of a threshold module: 1—thyristor, type BT151-800L; 2—Zener diode, type 1N5383; 3—two terminal boards, type Wieland 9700A/6S35.

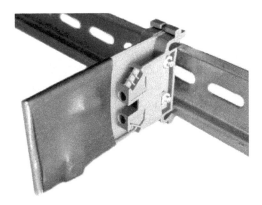

FIGURE 8.3
Single module installed on a standard DIN rail.

anode–cathode. Voltage drop on the open thyristor will not exceed a fraction of a volt, which, combined with the low current passing through it (15–50 mA), ensures insignificant power dissipation and does not cause heating of the thyristor.

Both elements are installed on a small plate made of laminated fiberglass (dimensions of 70 × 40 mm), clamped with two screws between a standard terminal block designed to be installed on a standard DIN (Deutsch industrial norm) rail. After the elements are mounted on the plate, they are covered with a layer of waterproof lacquer and insulated by means of a heat-shrinking dielectric tube that provides mechanical protection of elements and connections on the plate. A completed device represents a small, easily mountable module (Figure 8.3) at a low price. (A thyristor with a Zener diode costs less than USD 2.)

The module is not connected to all logic inputs, but rather only to the most critical, the actuation of which results in relay actuation and trips of power transmission lines and power equipment. One DPR may require two to four modules.

This device can be used in combination with an ordinary electromechanical relay in cases when the actual voltage of actuation is less than half of the DC circuit's voltage—in other words, when there is a danger of faulty actuation.

It should be noted that application of the suggested device does not actually reduce reliability of relay protection, since electronic elements are fully powerless when a control terminal is open and cannot actuate themselves (e.g., as a result of thyristor short-circuiting or self-opening). If we admit (theoretically) the possibility of short-circuiting of a module's electronic elements, relay protection will return in the operation mode, which existed before the application of the module. Only the breakage of an internal circuit of the module can affect relay protection reliability. However, statistical data show that internal breakages in semiconductor devices emerge only under very high rates of passing current (when the semiconductor structure burns out) and they account for less than 5% of all damages.

Under real current rates passing through the thyristor in the module's layout (15–50 mA), burnout and breakage of a thyristor's structure is impossible. Several internal soldered joints and an additional joint of conductors with a terminal block used in the module can really affect reliability of relay protection. This is why a printed circuit that can contain hidden defects cannot be used for mounting elements, while flexible terminals of the thyristor and Zener diode are fixed to each other before soldering by winding of thin copper tin coated wire (as used in some military applications); this reduces the probability of module failure to the smallest minimum.

8.2 Insulation Problems in Substation DCAPS

Substation and power plant DC auxiliary power systems (DCAPSs) are the most important components of power supply systems, without which the reliable operation of relay protection, automatic devices, and control systems is unrealizable. These circuits are fully insulated from the "ground" and are very long. In case of circuit insulation breakage on one of the terminal poles and connection of the ground to them, the DCAPS continues functioning normally, but this breakage should be revealed (through an alarm) and repaired as soon as possible, for example, in case of breakage of circuit insulation of additional and the second pole terminals, a short circuit will originate in the auxiliary power supply system. This is why branched DCAPSs should include devices to control insulation that measure resistance of insulation of both poles DC and issue an alert when insulation decreases lower than the set rate. However, revealing insulation problems is not enough; it is necessary to find the location of the breakage, which is very difficult without special tools.

Different Russian (such as IPI-1M, EKRA-SKI, SKIF, BIM P30, and others) and other foreign tools (Figure 8.4) that satisfy these requirements and help to find the breakage are used in practice. The leadership in the field of development and production of tools to control insulation belongs to the German company Walter Bender GmbH. As a rule, all these tools include a generator of test impulses, a receiver of these impulses, and a clamp meter.

Unfortunately, technical specifications of most types of tools do not contain the test current rate, which is generated by these tools in the circuit of the auxiliary DC power supply. The exception is represented by the BGL tool (Figure 8.5), manufactured by Multi-Amp; its maximum current rate is 110 mA. This current rate in the BGL tool as well as full negligence of ratings of operational current by manufacturers of other tools indicates that this

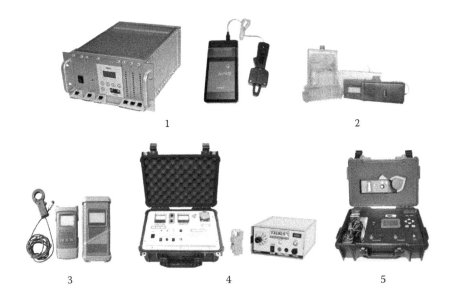

FIGURE 8.4
Some types of tools to search for short circuit to the "ground" in auxiliary DC circuits: 1—SKIF (Techelectro ST); 2—IPI-1M (Pskov electro-technical plant); 3—GAO A0E10003 (GAO Tek Inc.); 4—Digitrace DC (Taurus Powertronics Ltd); 5—GFL-1000 (Eagle Eye Power Solutions).

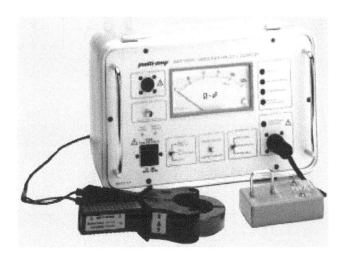

FIGURE 8.5
A device for searching location of a short circuit to the ground in auxiliary DC circuits, type BGL (Multi-Amp).

parameter is underestimated. This is a very important parameter, which can be crucial in selection of one or another type of tool for practical application.

Neglect of this parameter can result in serious accidents in the energy system while using the tool. For example, there was a case when switches were actuated and the voltage transmission line, together with the transformers, was cut off during searching for insulation breakage in the DC operational circuit by means of mobile tool EDS3065 manufactured by Bender.

What happens in DCAPS when the tool searching for broken insulation is working? Let us discuss this question taking the tools of Bender Company as an example (Figure 8.6).

When there is a short circuit of one DC pole to the ground in the DCAPS, the current through the short-circuit location will be stipulated by a leakage through normal insulation of the other (not damaged) pole only. It can be

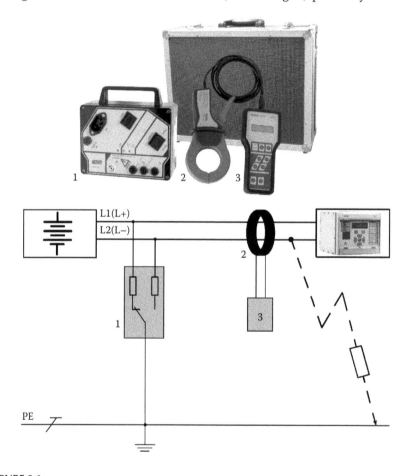

FIGURE 8.6
Principle of work of mobile tools searching insulation breakage in a DCAPS manufactured by BENDER.

vanishingly small, so it is very difficult to reveal the breakage and far more difficult to find its location. In order to increase the current, flow through the damaged place in the module (Figure 8.6) turns on additional adjustable resistance between each of the pole terminals and the ground automatically at a very low frequency (from several hertz to fractions of a hertz, depending on the tool). This limits current flow through the broken place, depending on voltage level of the DCAPS.

When there is no breakage, current rates flowing through a clamp meter in both directions are equal and thus compensate each other. If additional resistance is turned on and there is an insulation breakage, a differential current appears, the rate of which is a function of this resistance. It is clear that the higher the rate of this current is, the easier it is to find where the insulation is broken. As mentioned earlier, specifications of many tools do not mention the rate of this current at all, while the older tools of Bender had a switch that allowed selecting the current rate in a range from 10 to 25 mA.

What is so dangerous about the flow of such low current rates in the DCAPS? The problem is revealed when the breakage of insulation is located on the L section, which connects electromagnetic auxiliary relays or logic inputs of DPR with a switching-on of terminal K of the external control apparatus (Figure 8.5). In this case, when looking for a damaged place, the operational current of the tool will be flowing through the coil of the auxiliary relay through the logic input of DPR while the K terminal is open (Figure 8.7).

If the rate of the operational current of the tool is higher or even equal to current rate, which activates logic input of the DPR or results in actuation of

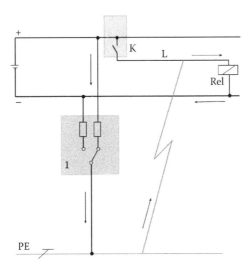

FIGURE 8.7
A diagram of circuit formation, which stipulates faulty actuation of auxiliary relays or activation of logic inputs of the DPR.

TABLE 8.1

Actuation (Pickup) Current Rates of Different Widespread Types of Auxiliary Electromagnetic Relays for 220 V DC with a Shunting Resistor

Pickup Current of Auxiliary Relay, mA	Resistance of a Shunting Resistor, kΩ
2.4	—
12.9	10
15.2	6.8
20.1	4.7

an auxiliary relay, Rel, be prepared for problems. The measurements show that the pickup current (do not confuse this with nominal operating current) of electromagnetic relays with high-resistance coils at 230 V DC amounts to 2.4 mA, whereas actuation current of logic inputs of the DPR, type REL/REC/RET 316 series (manufactured by ABB), amounts to 3–4 mA. Thus, when using most of the devices (which have 5 mA operating currents and higher) to find insulation damage in DCAPS, there is a high probability of faulty actuation and disconnection of high-voltage circuit breakers.

What should we do in this situation? First of all, it is necessary to find out what the operating current of the device is. If this current is higher than 1 mA (most devices), stop using this device immediately. Alternatively, one can buy a new device, such as EDS3091 (Bender) that works under currents of 1 and 2.5 mA. The price of this device is about USD 9,000. Another alternative is the artificial desensitization of some auxiliary relays and some logic inputs of DPR, faulty actuation of which can result in emergency turning off of the circuit breaker (CB). In order to do this, it is enough to shunt auxiliary relays and logic inputs of the DPR by resistors so that their actuation currents are higher than 10 mA. Table 8.1 shows actuation (pickup) current rates of auxiliary electromagnetic relays with nominal 220 V DC with coils (resistance of 19 kΩ) shunted by resistors with different resistance rates.

It is recommended to use 50 W wire resistors, types HSA50, THS50, WH50, and others, as shunting resistors in housings, which are convenient for mounting on the inner wall of metal cabinets (Figure 8.8).

The power that is constantly dissipated by resistors (if this input of a DPR or a coil of auxiliary relay is continuously underpowered) does not exceed 10 W (to ensure actuation current rates of 12–15 mA). This is why heating of 50 W resistors is insignificant, especially if they are mounted on the inner wall of the metal cabinet. When such resistors are installed, safe work on finding insulation damages in the DCAPS is ensured and there is no need to buy new devices.

Thus, we can conclude that mobile devices used to find insulation damages in the DCAPS are potentially dangerous for the power system, since their activity can cause the accidental turning off of CBs. It is absolutely necessary not to

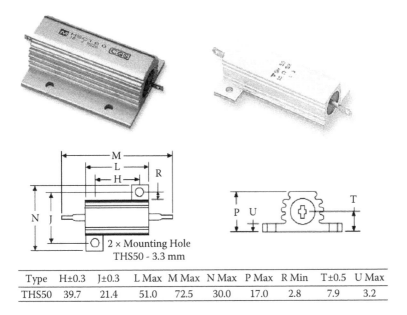

Type	H±0.3	J±0.3	L Max	M Max	N Max	P Max	R Min	T±0.5	U Max
THS50	39.7	21.4	51.0	72.5	30.0	17.0	2.8	7.9	3.2

FIGURE 8.8
Wire resistors (50 W) with convenient mounting on the inner wall of metal cabinets.

use devices with operational current rates exceeding 1–2 mA. They should be replaced with new generation devices that have reduced operational current rates, or some auxiliary electromagnetic relays and some logic inputs of the DPR, which are functionally related to turning off CBs, should be desensitized.

Another problem, which was detected when using devices to control insulation conditions in DCAPSs, is related to stationary equipment, which is used at substations with different voltage levels (e.g., 220 and 60 V [or 48 V]). When there are two systems of auxiliary DC power supply, stationary devices to control insulation conditions are mounted in the circuits of each of them. Additionally, a third device controls insulation conditions between these two systems (Figure 8.9).

When using IRDH375 devices, it was found that when the ground appears in a 220 V system, not only the B1 device is actuated, but also the B3. Furthermore, all the devices are synchronized with each other via their own communication channels so that when one device creates a test impulse of current, all other devices do not react to the ground. We could not determine the reasons for the faulty actuation of B3 even after discussion with a manufacturer, but this actuation makes it difficult to detect a damaged system at a substation. We are talking about visual detection of devices' conditions (i.e., actuated–not actuated), but not about output signal, which is combined from all three devices into one aggregate signal concerning insulation damage. This problem can be solved by means of three signal lights connected in a simple layout (Figure 8.10).

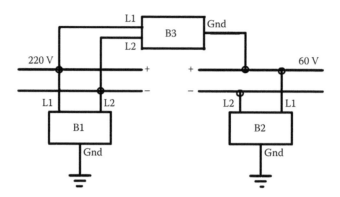

FIGURE 8.9
Connection diagram of stationary devices (B1, B2, B3) for insulation supervision on substation with two separate systems of auxiliary DC power supply with different voltage rates.

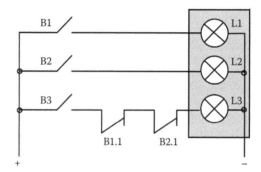

FIGURE 8.10
Indication circuit of insulation control device's condition, which does not react to faulty actuation.

The next problem is probable, since it has not really been detected as yet [2]. But there is a great probability that it will show up. We are talking about using stationary devices for insulation control, which consist of two separate systems, DC1 and DC2, with a backup of consumers by means of diodes, in DCAPSs (Figure 8.11). There is a big variety of stationary devices to control insulation conditions in DCAPSs; they implement different principles of operation and are manufactured by small companies in India, China, Canada, and the United States, as well as commercial giants, such as ABB and Siemens. The most widely used devices are those that control the balance between positive and negative potentials in relation to an artificially created grounded point in the device with a zero potential as well as those that create test pulses of current (similar to those mobile devices mentioned earlier) by occasional connection of "plus" and "minus" to the ground through a current limiting resistance.

If devices of the latter type are implemented in the layout indicated in Figure 8.11, there can be a situation when one device will close the minus of one bus system (B1) to the ground, and the other will close the plus of the second bus system (B2) to the ground. Moreover, both bus systems will be connected in series and consumers will receive double voltage (i.e., about 460 V). The current rate of 10–20 mA limited by the devices is sufficient to damage varistors and resistors at input circuits of I/O modules and power supply of DPRs. Indeed, cases of voltage summation from two bus systems in the layout with splitting diodes are known (because of other reasons) in DCAPS application practice, and they always result in massive damages of DPR power supplies. It is possible to prevent such a dangerous mode by using insulation control devices synchronized with each other when their simultaneous operation is eliminated or when using devices that do not implement operational principles based on current impulse creation.

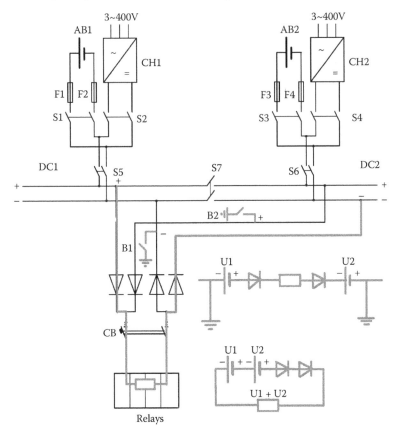

FIGURE 8.11
A diagram of creation of a circuit with double voltage in a DCAPS with two bus systems—DC1 and DC2—and splitting diodes in the load circuit.

In conclusion, it should be noted that the insulation control devices in the DCAPS are potentially dangerous and can result in the damage of electronic equipment and severe accidents in power networks. This is why it is imperative to be very careful when selecting them to study their specifications considering the previously mentioned recommendations.

References

1. Gurevich, V. I. 2008. Increase of noise immunity of logic inputs of microprocessor relay protection. *Electronica-Info* 11:26–27.
2. Gurevich, V. I. 2012. The problems of increase of reliability of RPA DC power supply systems. *Electric Energy Transmission and Distribution* 3:70–73.

9

Voltage Disturbances in
Auxiliary Power Supply

9.1 Electromagnetic Disturbances in the Power Network

Many cases of malfunctioning of and even damages to microprocessors are caused as a result of the impact of electromagnetic disturbances (*blackouts, noise, sags, spikes, surges*) from the power supply network on operation of the digital protective relays (DPRs).

9.1.1 Blackout

A blackout results in a total loss of utility power:

- *Cause:* Blackouts are caused by excessive demand on the power network, lightning storms, ice on power lines, car accidents, construction equipment, earthquakes, and other catastrophes.
- *Effect:* Current work in RAM or cache is lost and there is total loss of data stored on ROM.

9.1.2 Noise

More technically referred to as electromagnetic interference (EMI) and radio frequency interference (RFI), electrical noise disrupts the smooth sine wave one expects from utility power:

- *Cause:* Electrical noise is caused by many factors and phenomena, including lightning, load switching, generators, radio transmitters, and industrial equipment. It may be intermittent or chronic.
- *Effect:* Noise introduces malfunctions and errors into executable programs and data files.

9.1.3 Sag

Also known as brownouts, sags are short-term decreases in voltage levels. This is the most common power problem, accounting for 87% of all power disturbances according to a study by Bell Labs:

- *Cause:* Sags are usually caused by the start-up power demands of many electrical devices (including motors, compressors, elevators, and shop tools). Electric companies use sags to cope with extraordinary power demands. In a procedure known as rolling brownouts, the utility will systematically lower voltage levels in certain areas for hours or days at a time. Hot summer days, when air conditioning requirements are at their peak, will often prompt rolling brownouts.
- *Effect:* A sag can starve a microprocessor of the power it needs to function and can cause frozen keyboards and unexpected system crashes, which result in lost or corrupted data. Sags also reduce the efficiency and life span of electrical equipment.

9.1.4 Spike

Also referred to as an impulse, a spike is an instantaneous, dramatic increase in voltage. A spike can enter electronic equipment through AC, network, serial, or communication lines and damage or destroy components:

- *Cause:* Spikes are typically caused by a nearby lightning strike. Spikes can also occur when utility power comes back online after having been knocked out in a storm or as the result of a car accident.
- *Effect:* Catastrophic damage to hardware occurs. Data will be lost.

9.1.5 Surge

A surge is a short-term increase in voltage, typically lasting at least 1/120 of a second:

- *Cause:* Surges result from the presence of high-powered electrical motors, such as air conditioners. When this equipment is switched off, the extra voltage is dissipated through the power line.
- *Effect:* Microprocessors and similarly sensitive electronic devices are designed to receive power within a certain voltage range. Anything outside expected peak and RMS (root-mean-square; considered the average voltage) levels will stress delicate components and cause premature failure.

Malfunctions and damages of DPRs due to electromagnetic disturbances are described in the literature. For example, mass malfunctions of microprocessor-based time relays occurred in nuclear power plants in the United States [1]. A review of these events indicated that the DPR failed as a result of voltage spikes that were generated by the auxiliary relay coil controlled by the DPR. The voltage spikes, also referred to as "inductive kicks," were generated when the DPR contacts interrupted the current to the auxiliary relay coil. These spikes then arced across the contacts of the output relay of the DPR. This arcing, in conjunction with the inductance and wiring capacitance, generated fast electrical noise transients called "arc showering" (electromagnetic interference). The peak voltage noise transient changed as a function of the breakdown voltage of the contact gap, which changed as the contacts moved apart and/or bounced. These noise transients caused the microprocessor in the DPR to fail.

9.2 Voltage Sags in AC Networks

According to IEC 61000-4 [4,5] voltage sags (sometimes referred to as "dips") are brief reductions in voltage (below 0.8 U_N), lasting from tens of milliseconds to 15 s (see Figure 9.1). As is known, the main reason for voltage sags in the 0.4 kV network of substation auxiliary supplies is short circuits in external high-voltage grids. In manufacturing plants, such voltage sags are frequently associated with the working modes of the powerful electrical equipment—for example, with starting power motors. Voltage sags are an important criterion of the power quality.

9.2.1 Voltage Sags in Manufacturing Plant 0.4 kV Networks

Sags in the 0.4 kV networks of manufacturing plants may seriously disturb the production cycle due to the mass disconnection of electrical motors through the releasing of the contactors followed by the autostarting of a number of motors, which may cause additional serious decreasing in the voltage level and exacerbate the problem [6,7].

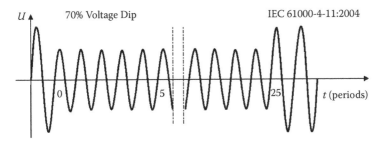

FIGURE 9.1
Example (from IEC 61000-4-11) of the 70% voltage sag during 25 cycles (0.5 s).

As shown in Fishman [8], upon voltage cessation for an electromotor during 0.4–0.8 s, the vectors of the residual electromotive force of the motor can occur in antiphase with the vector of the network voltage. As a result, upon network voltage restoration, the high magnitude current pulse will flow through the motor, causing the tripping of the protective circuit breaker and the disconnection of the motor.

On the other hand, short voltage sags with durations of less than 200–300 ms (most frequent in a 0.4 kV network) do not harm the motors. For these reasons, the means for ameliorating sags in networks in manufacturing plants have included, as usual, some technical solutions for keeping the contactors closed during a sag: special dynamic voltage sag compensators, uninterruptable power sources (UPSs), etc. Because such compensators and high-power UPSs are very expensive, different electronic devices [9,10] have been developed that guarantee feeding the contactor's coil from a DC power supply during short sags.

As is known, in the pickup process of an AC contactor there is a considerable variation of the current consumed by its coil, thereby leading to considerable variation in the core's attractive forces needed for contactor pickups. When feeding a contactor's coil from a DC supply, such current variations (and, as result, also the attractive force variations) will not be present and the contactor does not work properly.

These previously mentioned electronic devices use four levels of DC voltages for feeding the contactor coil, which simulates the natural attraction force characteristics at contactor pickups on AC. These electronic devices with integrated circuits (microchips) are not intended, however, for use with powerful contactors with low-resistance coils (10–15 Ω) and high-inrush currents. For example, the power consumed by the coil of the 3NTF54 contactor at pickup is 1.6 kVA on AC and 1.2 kW on DC (with special starting coil).

For large contactors with powerful coils, special devices have been developed that work on another principle (Figure 9.2). The device consists of undervoltage relay KU, timer K1 for an impulse-ON standard function, and a simple DC power supply consisting of transformer T, rectifier bridge VD2, and low-voltage, high-capacity capacitor C. When control switch S1 is closed, the 230 VAC (volt-ampere characteristic) is applied to the voltage relay KU, which picks up if the applied voltage is more than a minimum allowed value (180 V in our case) and closes its contact in the feeding circuit of the KT timer. The timer instantly picks up and by its contact connects the contactor's coil to the 230 VAC network through rectifier bridge VD1 and limiting resistor R1.

The direct current near 5 A will be carried through the contactor's coil. Such a current produces an electromagnetic force equivalent to that of the natural starting current, leading to the usual connection of the contactor to the 230 VAC network. Simultaneously, capacitor C1 is quickly charged. Due to the presence of diode VD3, capacitor C1 is charged from a low-voltage

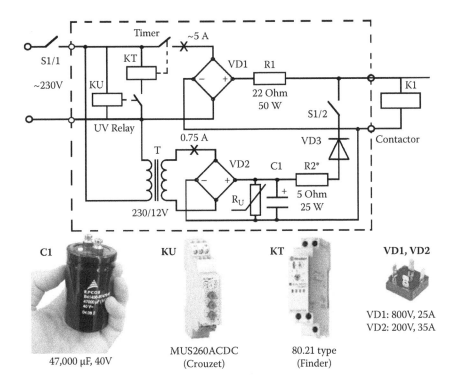

FIGURE 9.2
Circuit diagram of the control device for a powerful AC contactor.

(12 V) DC power supply only and is completely isolated from the high voltage (near 70 V) applied to the contactor's coil during the starting process. After 2–3 s, following the contactor pickups (the time is determined by the timer's internal setting), the timer's contact breaks off a high starting current in the contactor's coil. Thus, diode VD3 will instantly unlock and the low-voltage power supply with the charged capacitor C1 will connect to the contactor's coil. At this moment of time, the contactor's coil is fed by the lowered DC current—limited, in addition, by the low-resistance resistor R2.

Selection of the value of this resistor depends on the specific contactor type. For instance, for a 3TF54 type contactor, the resistance of this resistor has to be as low as 5 Ω. At this resistance, the reliable holding of the contactor in the closed position is provided over a long period of the AC voltage decreasing (down to 140–130 V) and, at the same time, the allowable temperature of heating of the coil (not to exceed 50°C–60°C) is provided.

The research that has been done has shown that when feeding the contactor's coil with the lowered DC current, its sensitivity to decreasing power supply voltage level is sharply reduced. For example, the contactor that was tested was held in the closed position at a voltage reduction on the coil from

12 down to 2–3 V—that is in four to six times more. This positive property is used in this device for holding the contactor for short-term downturns of the voltage level in AC networks. For very deep voltage sags or even full voltage loss, holding the contactor is affected by the energy of capacitor C1. From the results of the tests, it appears that small size capacitors with capacities in the range of 47,000 μF on 40 V are capable of holding the large contactor (in our case, the 3TF54) for periods of 1.3–1.5 s. This is quite sufficient for short-term voltage sags in real life in the AC networks.

The rectifier bridge VD needs to be selected with considerable reserve in connection with current because high current pulses flow through it during capacitor charging. At the fall of the voltage in AC network to a level lower than 160 V, the undervoltage relay KU opens its contact and disconnects the feeding circuit of the timer KT. However, the position of the output contact of the timer does not change, and the capacitor continues feeding the contactor's coil from the low-voltage DC power supply until the restoration of the proper voltage level in the AC network or until the capacitor C1 energy is fully exhausted (which will occur if the voltage sag lasts over a long time interval). At that time, the contactor will be disconnected. With the restoration of voltage in a network up to a level not less than 180 V, the undervoltage relay KU again will pick up, the timer feeding will appear, and the working cycle of the device as described earlier will repeat.

The impulse-ON function—sometimes called "interval," "fleeting," "single shot," "power ON," "single shot leading edge," or "rising edge pulse"— is not something exotic; it represents the standard function (designated, sometimes, as function number 21). The timers actualizing such functions are widely available in the market. These are, for example, timers of series PBO (Meander, Russia); CT-VWE, CT-WBS, CT-VWD (ABB); BC7931 (Dold & Soehne); MICV, NMICV (General Electric); KRDI (ABB); 3RP15 (Tyco Electronics); DIL-ET-11-30-A (Moeller); DDT, TZ (Tempatron); 87.21, 81.01 (Finder); MURc3 (Crouzet); RE7-PR11 (Telemecanique); M1SMT (Broyce Control); 3RP1505 (Siemens); TDRPRO-5100 (Magnecraft), and many others. Unfortunately, only some of them, such as 81.01, 80.01, and 80.21 (Finder), 821 (Magnecraft), 4604 (Artizan) and some others equipped with powerful output contacts, are suitable for switching coils of large contactors. Using timers of the other types will necessitate using an additional auxiliary relay with powerful contacts inserted in the circuit instead of the timer's contacts.

Voltage relay KU can be used as either an undervoltage relay with adjustable trip level or hysteresis, which does not required a separate power supply. Such relays as SUA145 (Bender), EUS (EID Electronics), MUS260ACDC (Crouzet), M200-V1U (Multitek), RM4-UB3 (Telemecanique), PVE (Entrelec), UAWA (Thiim A/S), BQP1202 (Midland Jay), PKH-1-1-15 (Meander, Russia), and others suit these requirements.

The device is assembled in a closed plastic container with dimensions of 210 × 160 × 90 mm. It is abundantly clear that the device can be used with

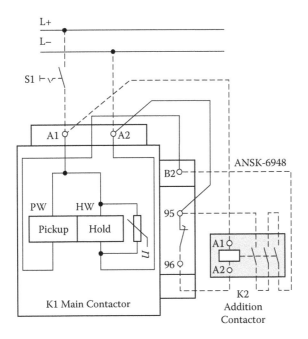

FIGURE 9.3
Solution offered by Siemens Company.

medium size contactors as well as with small contactors. In both these cases, the capacitance of the capacitor and the power of the transformer (and therefore its cost and dimensions) will dramatically decrease.

It needs to be noted that some manufacturers (including Siemens—manufacturer of the powerful contactors 3TF series) provide the possibility for feeding AC contactors from a DC power supply (DC network with power battery). In this case, the contactor's condition is fully independent of sags in the AC network. This offers one more way of solving the problem, but, on the other hand, realizing this solution is not simple because of special starting characteristics which need to be simulated as previously mentioned. Siemens offers two special windings for 3TF5 series contactors: a powerful pickup winding (PW) and a low-power holding winding (HW) (Figure 9.3).

Changeover from one winding to another after contactor K1 pickups is effected with the help of an additional contactor, K2, connected in series with powerful contacts (for disconnection of the high inductive load at 230 VDC [voltage dip compensator]) and an additional auxiliary contact block (95, 96) on the main contactor.

With the presence of a powerful battery and the possibility of setting the DC voltage to the contactor's location, the problem can be solved by means of a more intelligent method: as has already been mentioned, the timer 81.01 (Finder) type and a small switching type power supply for output of 12 V and

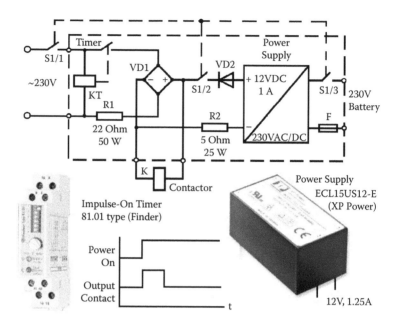

FIGURE 9.4
Device for feeding a large AC contactor from a DC auxiliary power supply.

1.2 A (Figure 9.4). Only two not very expensive off-the-shelf devices are needed for realizing the solution. The cost of these set elements is only about $120.

9.2.2 Voltage Sags in 0.4 kV Auxiliary AC Network

The peculiarity of the low-voltage auxiliary AC network in power substations is that it does not contain devices that allow for short pauses in the power supply and almost all of the critical power consumers (relay protection, emergency modes recorders, communication system, signaling, and remote control) are fed, as usual, from a power substation battery. At the same time, the power electronic systems with microprocessor controllers such as invertors, battery chargers, and power supplies are fed from the auxiliary AC network. Practical experience has shown that such devices do not "love" short voltage interruptions (50–200 ms) with subsequent voltage restoration. Sometimes, such devices have time to hang through automatic changeover from main to spare power supply (transformer).

Another problem of the power battery chargers with powerful input transformer is the high inrush current at sudden interruption and subsequent input voltage restoration that causes full charger disconnection by the electromagnetic releaser of the input circuit breaker. This state of affairs is considerably aggravated in some cases when even single-voltage sags with durations of 100–200 ms provoke multiple pickups and releases of the electromagnetic contactor during the sags.

9.3 Problematic Action of the Powerful Contactors as Changeover from Main to Reserved Auxiliary AC Power Supply on Substations

For increasing the reliability of an AC network of 0.4 kV in power substations, ordinarily two auxiliary transformers, feedings from different lines, are used. One of them connects to the 0.4 kV network permanently and the other one automatically connects at voltage disappearance on the first transformer. Connecting and disconnecting an AC network of 0.4 kV to these transformers is affected by means of two powerful electromagnetic contactors on 200–400 A with AC control coils. These contactors are the major elements of the auxiliary network on which, in many respects, reliable work of all power substations depends.

As an object of research, the electromagnetic contactor 3TF54 type (Siemens) with switching capability of 300 A has been taken (Figure 9.5), which is used at changeover in the auxiliary AC network of 0.4 kV in power substations. During the research, the oscillograms of pickups and releases of the contactor have been recorded at the feeding of the control coil from the AC supply (Figures 9.5 and 9.6). The oscillogram in Figure 9.5 shows the presence of high starting (inrush) current caused by small impedance of the coil until the moment of the closing of the contactor's magnetic circuit. Oscillograms shown in Figure 9.6 allow determining the time pickups and time release of the contactor (i.e., the reaction time of the contactor for voltage sags).

Analysis of the oscillograms has shown that full switch-on time (i.e., the time interval between the moment that the voltage is applied to the coil and

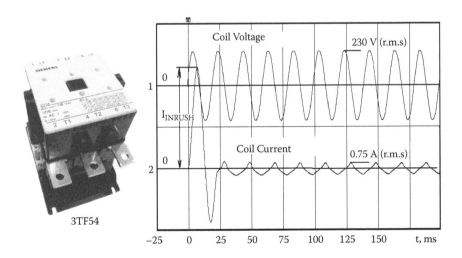

FIGURE 9.5
Electromagnetic AC contactor 3TF54 type (Siemens) and oscillograms for current and voltage on its coil at switch-on.

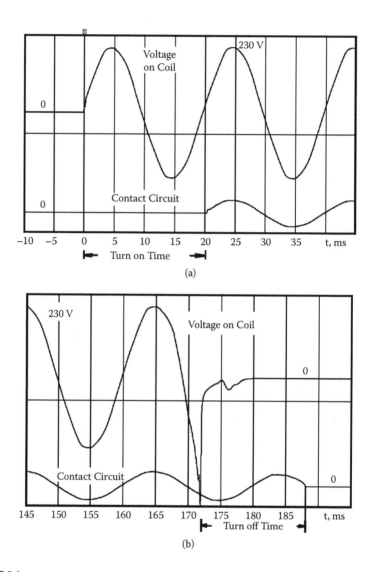

FIGURE 9.6
Oscillograms of contactor, type 3TF54, switching on and switching off.

the moment of main contacts closing) is about 20 ms. And the full switch-off time (i.e., the time interval between the disconnection of voltage on the coil and the moment of main contacts opening) is 15–18 ms.

The data sheet indicates 10–30 ms for nominal voltage applied before disconnection and 10–15 ms for voltage 0.8 of the nominal value. Such small time delays for such large contactors means that, during typical voltage disturbances with alternating voltage level sags and restorations, the contactor will have time to connect and disconnect the main power circuit several

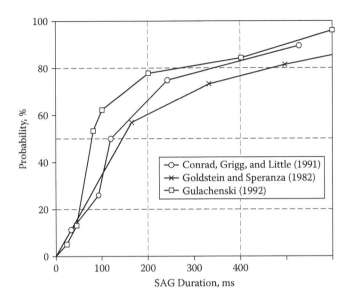

FIGURE 9.7
Sag duration in power AC network according to some researchers.

times. Moreover, as shown in Iyoda et al. [11], the reaction of the contactor during a 75% voltage sag is more serious than at 100%, since the releasing time for the first case is shorter by 40%–50% than in the second case and may be 10 ms even for a large apparatus.

One more problem of the power AC contactor was discovered during the research stage. It was found that reducing voltage across the coil up to 150–135 V causes high vibration of the contactor magnetic system with magnitude that is sufficient for closing and opening its main contacts. The same phenomenon arises when the AC voltage across the coil increases from 0 up to 160–185 V. The possibility of working in such a mode with such speed means that even at a single-voltage sag (to a level of 135–150 V during 100–200 ms; Figure 9.7), it transforms in the power generator multiple interruptions of the main voltage in the substation auxiliary AC network. The same result appears when trying connection of the contactor's coil to a power supply with 150–170 V.

9.3.1 Offered Solution for the Problem

In view of the character of the loads fed from the auxiliary AC network in substations (power electronic equipment sensitive to short sags), the technical solution offered for contactors intended for use in manufacturing plants (retention of the contactor in a closed position at short sags) is not the correct technical solution for a substation of 0.4 kV network because, through closed

contactors' contacts, the short sags will affect sensitive equipment and provoke disturbances.

In our opinion, the problem must be solved not by means of retention of
the contactor in the closed position at voltage sags, but rather by means of
the rapid (during 10–12 ms) disconnection of the contactor's coil at voltage
levels dropping below 160 V and returning it to initial condition at voltage
level restoration up to 185 V with a time delay 5–10 s. A single interruption of
5–10 s duration in the auxiliary 0.4 kV AC substation power network does not
cause serious disturbances in the substation equipment due to a power battery feeding most of the important substation consumers. At the same time,
such an algorithm of the contactor's work may prevent serious disturbances
and failures in AC power electronic equipment.

For fast contactor disconnection at voltage network drops, most electronic
relays available in the market are not suitable because their minimal reaction
time is usually 100 ms. During a time interval of this duration, the contactor
will make several disconnections and connections of the main power.

As a result of our research, only a few types of the devices suitable for
a contactor's control were found (Figure 9.8). One of them is an undervoltage relay combined with a timer: Brown-Out Timer GBP2150 type, manufactured by Midland Jay (division of Midland Automation, United Kingdom).
The reaction time of this device for voltage drops of 30% is only 5 ms. The
release time after voltage restoration up to 80% can be adjusted to intervals
of 1–10 s. Another good example is the Russian undervoltage relay PKH-1-3-
15 type, manufactured by ZAO Meander (St. Petersburg). Such devices are
ideal solutions for our purposes. For decreasing loads on the output contacts in these devices, an additional auxiliary fast electromagnetic relay, type
58.32.8.230 (Finder), with powerful output contacts and releasing time of 3
ms is employed.

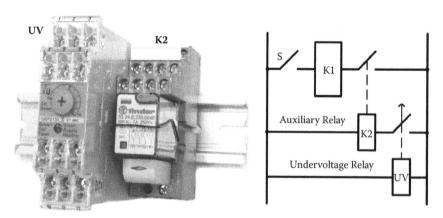

FIGURE 9.8
Device for fast forceful disconnection of the main contactor at voltage sag.

9.3.2 Conclusion

For manufacturing plants with electrical motors as dominant consumers and for power substations with power electronic equipment as dominant consumers in 0.4 kV AC networks, different methods must be used in contending with voltage sags. In the first case, the devices with capacitor or compact switching power supply for detaining the main contactor in a closed position during short voltage sags, as described earlier, can be used. In the second case, using a simple device for fast forceful disconnection of the main contactor at voltage sags below 25%–30% is recommended.

9.4 Problems in Auxiliary DC Power Supply

9.4.1 Problem of Power Supply of Relay Protection at Emergency Mode

As is known, both AC and DC voltages are used as auxiliary power supplies at substations. Use of DC auxiliary voltage increases the essential reliability of relay protection due to use of a powerful battery capable of supporting the required voltage level on the crucial elements of the substation at emergency mode with the AC power network disconnected. However, this increase of reliability comes at the cost of an essential rise in price of the substation and its maintenance. On the other hand, electromechanical relays of all types do not demand an external auxiliary power supply for proper operation as DPRs. Electromechanical relay operation requires input signals only (from voltage or current transformers). There may be some problem when it is necessary to energize the trip coil of the high-voltage circuit breaker by electromechanical protective relays at loss of auxiliary AC voltage in the emergency mode in substations with AC auxiliary power supply. However, this problem has been solved for a long time simply through use of a storage capacitor.

9.5 Voltage Dips in Auxiliary Power Supply Substation Systems

The problem of voltage dips in relay protection power supply circuits (AC, first of all) is as important as voltage dips in the auxiliary power plants and substations supplying their own needs. In the era of simple electromechanical relays, which did not require external power supplies (such as DPRs), the problem was limited to the necessity of supplying power to the trip coil

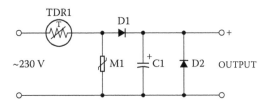

FIGURE 9.9
A typical layout of a capacitor unit the purpose of which is to supply power to the trip coil of a CB during voltage dips in the relay protection and auxiliary supply circuits.

of a circuit breaker (CB) during voltage dips. The problem was successfully solved 50 years ago using special units with reservoir capacitors, which supplied a necessary pulse of discharging current in the trip coil of a CB. The capacitor in this unit receives a constant charge from the AC circuit through a D1 diode (Figure 9.9).

In order to limit the capacitor's charging pulse current, which goes through a D1 diode during connection of the noncharged capacitor to a power supply, posistor TDR1 is used, while in order to protect the capacitor from spikes of overvoltage, a varistor M1 is used. Some manufacturers (e.g., Square D) add LED (light-emitting diode) indicators associated with the power to their circuits as well as threshold elements on thyristors, which ensure full charging of the capacitor until it is discharged to the load and used during multiple charging cycles (Figure 9.10).

For more complicated electromechanical protection, which includes many auxiliary relays in their internal circuits, power supply is necessary during voltage dips. The pulse accumulators of energy based on capacitors described previously cannot supply the power of such relays during the time necessary for their actuation, especially if there are time-delay relays.

As a result, complex power units appeared that supply power to relays simultaneously from a current transformer, a voltage transformer, and a capacitor (Figure 9.11). In the USSR, the first complex power units to supply power to relay protection were developed in the All-Union Research and

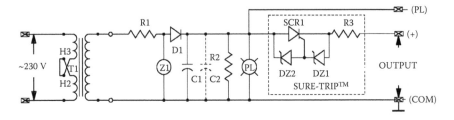

FIGURE 9.10
A complicated layout of a capacitor power unit Sure-Trip (Square D) with an additional threshold element on a thyristor SCR1, LED PL, and Zener diode Z1.

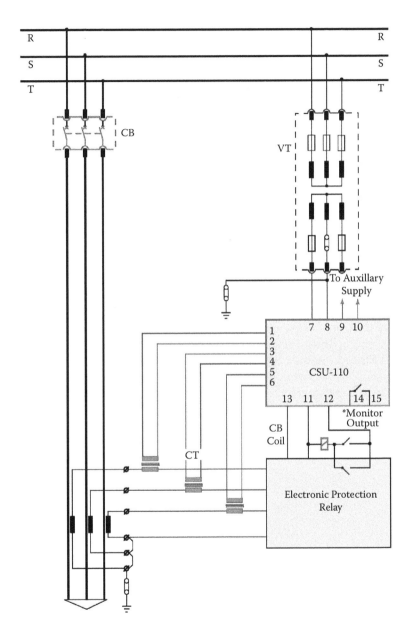

FIGURE 9.11
A connection diagram of a capacitor unit CSU-110 (Switching Systems Electronic Engineers) with a complex power supply from current transformers (CTs), voltage transformers (VTs), and the auxiliary power supply.

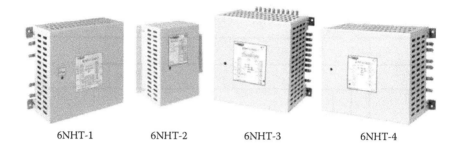

6NHT-1   6NHT-2   6NHT-3   6NHT-4

FIGURE 9.12
Complex power units manufactured by the Cheboksary Plant of Electric Apparatus.

the Scientific Institute of the Power Industry approximately simultaneously with the units of power reservoir capacitors. Today, these power units are manufactured by many companies including Cheboksary Plant of Electric Apparatus, Mechanotronica, and others (Figure 9.12).

Moreover, in order to ensure a more efficient use of the capacitor's capacity, it was charged to the voltage rate exceeding nominal voltage of a trip coil. For example, a capacitor in power units BPZ-400 with a block of reservoir capacitors BK-400 is charged to 400 V. These power units (Figure 9.13) are manufactured by the Cheboksary plant. Similar devices are manufactured by dozens of foreign companies, including leading electrotechnical companies, such as General Electric, Siemens, ABB, Alstom, and others (Figure 9.14).

Internal self-discharge of large-capacity electrolytic capacitors, which are used in capacitor power units, is limited in the time within which the

FIGURE 9.13
Power unit BPZ-400 manufactured by the Cheboksary Plant of Electric Apparatus that works in combination with the unit of reservoir capacitors BK-400.

FIGURE 9.14
Complex power units of foreign manufacturers.

capacitor saves energy sufficient for reliably tripping a CB. This is why built-in Ni-Cd baby batteries are used to charge capacitors in the most advanced models of capacitor blocks, when all external power supply options are lost. These baby batteries are constantly charging under normal regime, but when external power is lost, they supply power to a built-in low-power converter, which transforms the low voltage of the baby battery into the high voltage necessary to charge a capacitor.

The energy of baby batteries in these units (Figure 9.15) is enough to maintain a capacitor fully charged for 72 hours, when all types of external power supply are fully lost. Such compact devices are issued by many companies and allow keeping the capacitor charged for several days. Clearly, in such conditions, sufficient reliability of relay protection, even on an auxiliary AC, is provided. For this reason, the auxiliary AC is applied widely.

The situation began to change with the introduction of DPRs and their mass replacement of electromechanical relays. To the many problems caused by this transition [12–14], one more problem was added. As is known, the internal switching-mode power supply, admitting use as auxiliary AC and DC voltages, has an overwhelming majority of digital protective relays. Therefore, at first sight, there should be no reason to interfere with the use of an auxiliary AC voltage on substations with DPR.

FIGURE 9.15
Capacitor power units with Ni-Cd baby batteries, maintaining capacitor's charge over 72 hours.

The problem arises when there is not enough power for normal operation of an overwhelming majority of DPRs and only the presence of corresponding input signals (as for electromechanical relays); this also requires a feed from an auxiliary supply. How will the DPR behave at the loss of this feed at failure mode when the hard work of the microprocessor and other internal elements is required? How will the complex relay protection (containing some of the DPR incorporated in the common system by means of the network communication when there are also losses of auxiliary feed) function? How will the DPR behave during voltage sags (brief reductions in voltage, typically lasting from a cycle to a second or so, or tens of milliseconds to hundreds of milliseconds) during failure? We shall try to understand these questions.

The internal switching-mode power supply of the DPR contains, as a rule, a smoothing capacitor of rather large capacity that is capable of supporting the function of the relay during a short time period. According to research led by General Electric [15] for various types of DPRs, this time interval takes 30–100 ms. (However, for modern DPRs of a new generation the situation is different; see Table 9.1.) In view of the time of reaction, the DPR for emergency operation lies in the same interval and depends on that type of emergency mode; it is impossible to tell definitely whether protection will have sufficient time to work properly. At any rate, it is not possible to guarantee its reliable work. It is an especially problematic functioning of protection relays

TABLE 9.1

Withstanding of Voltage Interruption by Some Modern DPRs

DPR Type and Manufacturer	Maximal Power Interruption Duration without Disturbances in Relay Functions	Minimal Level of Voltage Supply Needed for Proper Functioning of DPR with 230 V Nominal Voltage
SIPROTEC 7UT6135 Siemens	1.6	78
SIPROTEC 7UT6125 Siemens	1.6	36
SIPROTEC 7SJ8032 Siemens	3.8	44
T60 General Electric	—	80
P132 Areva	—	45

with the time delay—for example, the distance protection with several zones (steps of time delay, reaching up to 0.5–1.0 s and more). Also, it is only possible to guess what will take place with the differential protection containing two remote complete sets of the relay at loss of a feed of one of them only.

Voltage sags are the most common power disturbance. At a typical industrial site, it is not unusual to see several sags per year at the service entrance and far more at equipment terminals. These voltage sags can have many causes, among which may be peaks of magnetization currents, most often at inclusion of power transformers. Recessions and the rises of voltage sometimes at failures and in transient modes are especially dangerous when occurring successively with small intervals of time. The level and duration of sags depend on a number of external factors, such as capacity of the transformer, impedance of a power line, remoteness of the relay from the substation transformer, and the size of a cable through which feed circuits are executed.

DPRs also have a wide interval of characteristics on allowable voltage reductions. As mentioned in Gurevich [14], various types of DPRs keep working capability at auxiliary voltage reduction from the rated value of up to 70–180 V. Thus, a DPR with a rated voltage of 240 V supposes a greater (in percentage terms) voltage reduction than devices with a rated voltage of 120 V. It is also known that any microprocessor device demands a long time from the moment of applying a feed (auxiliary voltage) to full activation at normal mode. For a modern DPR with a built-in system of self-checking, this time can reach up to 30 s. This means that even after a short-term failure with auxiliary voltage (voltage sag) and subsequent restoring of voltage level, relay protection still will not function for a long time.

What is the solution to the problem offered by the experts [15] from General Electric? Noting that existing capacitor trip devices obviously are not sufficient to feed the DPR, as reserved energy in them has enough only for creation of a short-duration pulse of a current and absolutely not enough to feed the DPR, the author comes to the conclusion that it is necessary to use a UPS for feeding the DPR in an emergency mode. The second recommendation of the author—to add an additional blocking element (a timer, for example, or internal logic of the DPR)—will prevent closing of the circuit breaker before the DPR becomes completely activated. Both recommendations are quite legitimate. Here, only usage of a UPS with a built-in battery is well known as a solution for maintenance of a feed of crucial consumers in an emergency mode. This solution has obvious foibles and restrictions (both economic and technical). Use of blocking for switching on the circuit breakers can be a very useful idea that should undoubtedly be used; however, it does not always solve the problem as failures of voltage feeding connected to operation of the circuit breaker are always a possibility.

In our opinion, instead of UPS usage a more simple and reliable solution of the problem is the use of a special capacitor with large capacity connected in parallel to the feed circuit of every DPR. High-quality capacitors with large capacity and rated voltage of 450–500 V are sold today by many companies for less than approximately $250 and are not deficient (see Table 9.2).

Elementary calculation shows that, when charged up to 250 V, one 5000 µF capacitor is capable of feeding a load with consumption power of 30–70 VA up to voltage decreasing to a minimum level of 150 V during 3–5 s; this is quite enough for operation of the DPR in the emergency mode.

Use of such a capacitor for auxiliary voltage of 220 V AC requires, naturally, a rectifier and some more auxiliary elements (Figure 9.16). In this device a capacitor of large capacity is designated, such as C2. The C1 auxiliary nonelectrolytic capacitor with capacity in some microfarads serves for

TABLE 9.2

Parameters of Capacitors with Large Capacity and Rated Voltage of 450–500 V

Capacity (µF)	Rated Voltage (V)	Dimensions (Diameter × Height) (mm)	Manufacturer and Capacitor Type
6,000	450	75 × 220	EVOX-RIFA PEH200YX460BQ
4,700	450	90 × 146	BHC AEROVOX ALS30A472QP450
10,000	450	90 × 220	EVOX-RIFA PEH200YZ510TM
4,000	500	76.2 × 142	Mallory Dur-Cap 002-3052
4,000	450	76.2 × 142	CST-ARWIN HES402G450X5L
6,900	500	76.2 × 220	CST-ARWIN CGH692T500X8L

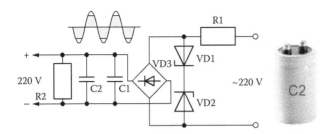

FIGURE 9.16
The device for reserve feed of a DPR at emergency mode with AC auxiliary voltage.

smoothing pulsations on electrolytic capacitor C2. It is possible to include also in parallel to C1 one more nonelectrolytic capacitor with a capacity of some microfarads, for protection of C2 against the high-frequency harmonics contained in main AC voltage. Resistor R1 (200–250 Ω) limits the charging current of C2 at a level near 1 A. The same resistor also limits pulse currents proceeding through back-to-back connected Zener diodes VD1 and VD2.

Resistor R2 has high resistance and serves to accelerate the discharging capacitor up to a safe voltage at switching off the auxiliary voltage. Zener diodes are intended for the maximal value voltage limits of capacitor C2 at a level of 240 V. Without such limitations on the device, output voltage would reach a value of more than 300 V due to the difference between RMS and peak values of voltage. That is undesirable both for DPR and for C2.

The Zener diodes slice part of a voltage sinusoid in which amplitude exceeds 240 V, forming a voltage trapeze before rectifying. As powerful Zeners for rating voltage above 200 V are not at present on the market, it is necessary to use two series-connected Zeners with dissipation power of 10 W and rating voltage of 120 V as each of the Zeners (VD1, DD2—for example, types 1N1810, 1N3008B, 1N2010, NTE 5223A).

As further research of this type of situation has clarified, the problem of maintenance of reliable feed DPR is relevant not only for substations with AC auxiliary voltage, but also for substations with DC voltage. Many situations where the main substation battery becomes switched off from the DC bus bars are known. In this case, nothing terrible occurs, as the voltage on the bus bar is supported by a charger. However, if during this period an emergency mode occurs in a power network, the situation appears to be no better, since use of an AC auxiliary voltage as charger feeds from the same AC network. Usually, an electrolytic capacitor with some hundreds of microfarads for smoothing voltage pulsations is included on the charger output. Since not only many DPRs but also sets of other consumers are connected to charger output, it is abundantly clear that this capacity is not capable of supporting the necessary voltage level on the bus bars during the time required for proper operation of the DPRs. Our research has shown that such high capacitance as 15,000 μF does not provide proper functioning of a DPR as consumption from a charger reaches up to 5–10 A.

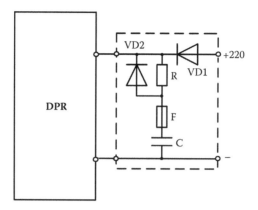

FIGURE 9.17
The device for reserve feed of a DPR at emergency mode with DC auxiliary voltage.

For maintenance of the working capability of DPRs in these conditions, it is possible to use the same technical solution with the individual storage capacitor connected in parallel to each DPR feeding circuit. Now the design of the device will be much easier, due to a cutout from the circuit diagram of Zeners and the rectifier bridge (Figure 9.17). The resistor R (100 Ω) is necessary for limiting the charging current of the capacitor at switching on auxiliary voltage with a fully discharged capacitor. Diode VD1 should be for a rated current of not less than 10 A. High-capability, quick-blow fuse F (5 A/1500 A, 500 V) is intended for protection of both the feeding circuit of the DPR and the external DC circuit when the capacitor is damaged.

The prototype of such a device with a capacitor of 3700 µF (Figure 9.18) has shown excellent results at tests with the various loadings that simulate DPRs of various types with different power consumption at nominal voltage of 240 V. American Schweitzer Engineering Laboratories (SEL) has changed its tack. It produces a complex power supply unit for the DPRs containing a small 48 V battery and charger (see Figure 9.19). Relay protection systems equipped with such devices can easily withstand prolonged voltage dips in auxiliary power supply or even long-lasting full blackouts.

One more variant of the solution of this problem for substations with DC auxiliary voltage is not to use an individual capacitor for each DPR, but rather a special "supercapacitor" capable of feeding a complete relay protection system set together with conjugate electronic equipment within several seconds.

Such supercapacitors can already be found on the market under names such as "supercapacitors," "ultracapacitors," "double-layer capacitors," and "ionistors" (in the Russian technical literature). There are electrochemical components intended for storage of electric energy. On specific capacity and speed of access to the reserved energy, they occupy an intermediate position between large electrolytic capacitors and standard accumulator batteries, differing from one another and the others in their principle of action, based

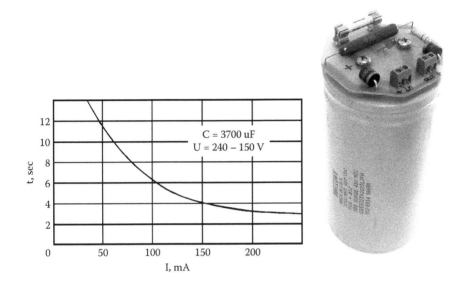

FIGURE 9.18
The prototype of the device for reserve feed of the DPR.

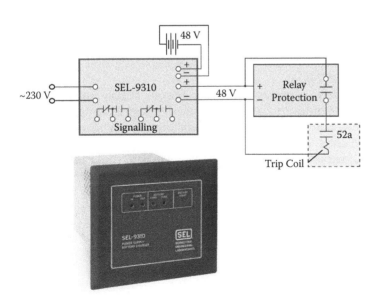

FIGURE 9.19
Complex power supply unit with charger and small 48 V battery manufactured by SEL.

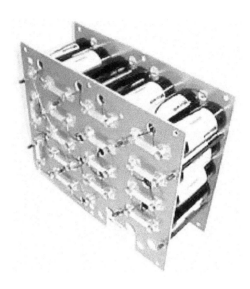

FIGURE 9.20
Internal design of high-voltage (10 voltages) supercapacitor, assembled from a number of low-voltage elements.

on redistribution of charges in electrolyte, and their concentration on the border between the electrode and electrolyte.

The capacity of modern supercapacitors reaches hundreds and even thousand of farads. Today, supercapacitors are produced by many Western companies (Maxwell Technologies, NessCap, Cooper Bussmann, Epcos, etc.) and also some Russian enterprises (ESMA, ELIT, etc.); however, the rated voltage of one element does not exceed, as a rule, 2.3–2.7 V. For higher voltage, separate elements connect among themselves in parallel and series as consistent units (Figure 9.20).

Unfortunately, supercapacitors are not so simply incorporated among themselves as ordinary capacitors; they demand leveling resistors at series cells' connection and special electronic circuits for alignment of currents at parallel cells' connection. As a result, such units turn out to be rather "weighty," expensive, and not so reliable. (There could be enough damage to one of the internal auxiliary elements to cause failure of the entire unit.) For example, a combined supercapacitor manufactured by the NessCap firm, with a capacity of 51 F and voltage of 340 V, weighs 384 kg! One unique company known to us that produces individual modules (i.e., not containing too many low-voltage cells inside) for high voltage (Figure 9.21) is the Canadian firm Tavrima. Its ESCap90/300 type supercapacitor (see Table 9.3) meets our purposes quite well. Another example is the supercapacitor module Sitras® series from Maxwell Technologies (Figure 9.22).

At the use of supercapacitor SC, the feeding circuit of the protective relays should be allocated into a separate line connected to the DC bus bar through

ESCap 90/300

FIGURE 9.21
High-voltage supercapacitors made as a single module and main parameters of the ESCap90/300 type capacitor.

TABLE 9.3

Main Parameters of ESCap90/300 Type Supercapacitor

Rated voltage, V	300
Capacitance, F	2.0
Max. power, kW	75
Max. energy, kJ (at 300 V)	90
Internal resistance, Ω	0.3
Dimensions, mm	Dia. 230×560
Weight, kg	35
Temperature, °C	–40 +55
Price per unit (for 2006), USD	1000.00

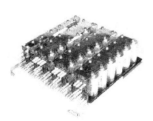

Sitras®
Nominal voltage –DC 750 V
Number of Ultracapacitors in single
Module – 1344
Energy stored – 2, 3 kWh
Energy saving per h – 65 kWh/h
Max. power – 1 MW
Capacitor efficiency – 0,95
Temperature domain – 20 to 40°C

FIGURE 9.22
High-voltage supercapacitor module Sitras series from Maxwell Technologies.

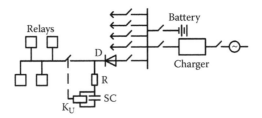

FIGURE 9.23
Example of usage of the supercapacitor as group power supply for protective relays at emergency mode with DC auxiliary voltage. K_U: voltage relay; SC: the supercapacitor.

diode D (Figure 9.23). Due to the large capacity of the supercapacitor, the voltage reduction on feeding input of a DPR at emergency mode (with loss of an external auxiliary voltage) will occur very slowly, even after passage of the lowest allowable limit of the feeding voltage.

From the personal experience of the author, cases of false operation of the microprocessor systems have been known to occur at slow feeding voltage reduction below allowable levels. This can be explained by the existence of different electronic components of a high degree of integration serving the microprocessor, having different allowable levels of voltage feeding reduction, and stopping the process of voltage reduction serially, thus breaking the internal logic of the DPR operation.

If such equipment is found in the DPR used on the given substation in parallel to the supercapacitor, it should be connected to a simple voltage monitoring relay K_U, which disconnects the supercapacitor at a voltage reduction below the lowest allowable level—for example, lower that 150–170 V.

When discussing the solutions of relay units' protection from voltage dips, it is noteworthy that there are special devices on the market that are meant to compensate voltage dips. These devices are called voltage dip compensators. There are several principles of building VDCs (Figure 9.24). One of them (Figure 9.24a) is very similar in principle to the VDCs uninterruptable power supply units and contains a battery, inverter, and a quick switch, which switches power to the inverter in the event of a voltage dip. This device enables compensating deep (up to zero) and lengthy voltage dips and does not require a series of accumulator batteries.

A much easier and cheaper option is a device that is part of a transformer with branches and quick semiconductor keys (Figure 9.24b). This device has a limited range of voltage dip compensation.

An intermediate position in terms of quality of compensation between those mentioned previously belongs to a device with a boost transformer, a reservoir capacitor, an inverter, and a quick key (Figure 9.24c).

Today, there are dozens of companies in the market that manufacture many models of special VDCs for any thickness of a container, from small low voltage (Figure 9.25) to cabinets with a voltage rate of hundreds of kilowatts (Figure 9.26).

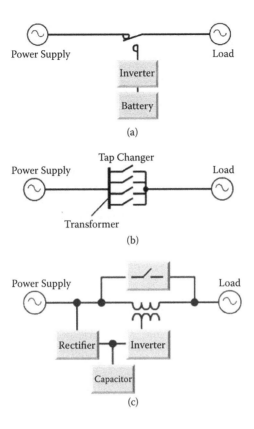

FIGURE 9.24
Principles of building the VDC.

FIGURE 9.25
Low-power VDC (up to several kilowatts).

FIGURE 9.26
High-capacity three-phase VDC (hundreds of kilowatts).

References

1. Information notice no. 94-20. 1994. Common-cause failures due to inadequate design control and dedication. Nuclear Regulatory Commission, March 17, 1994.
2. *The power protection handbook.* 1994. APC.
3. Dshochov, B. D. 1996. Features of power supply of computer network elements. *Industrial Power Engineering* 2:17–24, Rus.
4. IEC 61000-4-11 Ed. 2.0 b:2004. Electromagnetic compatibility (EMC)—Part 4-11: Testing and measurement techniques. Voltage dips, short interruptions and voltage variations immunity tests.
5. IEC 61000-4-34 Ed. 1.0 b:2005. Electromagnetic compatibility (EMC)—Part 4-34: Testing and measurement techniques—Voltage dips, short interruptions and voltage variations immunity tests for equipment with input current more than 16 A per phase.
6. Melhorn, C. J., T. D. Davis, and G. E. Beam. 1988. Voltage sags: Their impact on the utility and industrial customers. *IEEE Transactions on Industry Applications* 34 (3): 549–558.
7. McGranaghan, M. F., D. R. Mueller, and M. J. Samotyj. 1993. Voltage sags in industrial systems. *IEEE Transactions on Industry Applications* 29 (2): 397–404.
8. Fishman, V. 2004. Voltage sags in industrial networks. *Electrical Engineering News* 5 (29), and 6 (30), Rus.
9. Kelley, A., J. Cavaroc, J. Ledford, and L. Vassalli. 2000. Voltage regulator for contactor ride-through. *IEEE Transactions on Industry Applications* 36 (2): 697–703.
10. Andgara, P., G. Navarro, and J. I. Perat. 2007. A new power supply system for AC contactor ride-through. 9th International Conference Electric Power Quality and Utilization, Barcelona, October 9–11.

11. Iyoda, I., M. Hirata, N. Shigei, S. Pounyakhet, and K. Ota. 2007. Effect of voltage sags on electro-magnetic contactor. 9th International Conference Electric Power Quality and Utilization. Barcelona, October 9–11, 2007.
12. Gurevich, V. 2005. *Electrical relays: Principles and applications.* London: Taylor & Francis Group, 704 pp.
13. Gurevich, V. 2006. Nonconformance in electromechanical output relays of microprocessor-based protection devices under actual operation conditions. *Electrical Engineering & Electromechanics* 1:12–16.
14. Gurevich, V. 2006. Microprocessor protective relays: New perspectives or new problems? *Electrical Systems and Networks* 1:49–60.
15. Fox, G. H. 2005. Applying microprocessor-based protective relays in switchgear with AC control power. *IEEE Transactions on Industry Applications* 41 (6): 1436–1443.

Index

Milton Keynes UK
Ingram Content Group UK Ltd.
UKHW031147141024
449569UK00024B/1003